Annals of Mathematics Studies

Number 58

ANNALS OF MATHEMATICS STUDIES

Edited by Robert C. Gunning, John C. Moore, and Marston Morse

1. Algebraic Theory of Numbers, *by* HERMANN WEYL
3. Consistency of the Continuum Hypothesis, *by* KURT GÖDEL
11. Introduction to Nonlinear Mechanics, *by* N. KRYLOFF *and* N. BOGOLIUBOFF
20. Contributions to the Theory of Nonlinear Oscillations, Vol. I, *edited by* S. LEFSCHETZ
21. Functional Operators, Vol. 1, *by* JOHN VON NEUMANN
24. Contributions to the Theory of Games, Vol. I, *edited by* H. W. KUHN *and* A. W. TUCKER
25. Contributions to Fourier Analysis, *edited by* A. ZYGMUND, W. TRANSUE, M. MORSE, A. P. CALDERON, *and* S. BOCHNER
28. Contributions to the Theory of Games, Vol. II, *edited by* H. W. KUHN *and* A. W. TUCKER
30. Contributions to the Theory of Riemann Surfaces, *edited by* L. AHLFORS *et al.*
33. Contributions to the Theory of Partial Differential Equations, *edited by* L. BERS, S. BOCHNER, *and* F. JOHN
34. Automata Studies, *edited by* C. E. SHANNON *and* J. MCCARTHY
38. Linear Inequalities and Related Systems, *edited by* H. W. KUHN *and* A. W. TUCKER
39. Contributions to the Theory of Games, Vol. III, *edited by* M. DRESHER, A. W. TUCKER *and* P. WOLFE
40. Contributions to the Theory of Games, Vol. IV, *edited by* R. DUNCAN LUCE *and* A. W. TUCKER
41. Contributions to the Theory of Nonlinear Oscillations, Vol. IV, *edited by* S. LEFSCHETZ
42. Lectures on Fourier Integrals, *by* S. BOCHNER
43. Ramification Theoretic Methods in Algebraic Geometry, *by* S. ABHYANKAR
44. Stationary Processes and Prediction Theory, *by* H. FURSTENBERG
45. Contributions to the Theory of Nonlinear Oscillations, Vol. V, *edited by* L. CESARI, J. LASALLE, *and* S. LEFSCHETZ
46. Seminar on Transformation Groups, *by* A. BOREL *et al.*
47. Theory of Formal Systems, *by* R. SMULLYAN
48. Lectures on Modular Forms, *by* R. C. GUNNING
49. Composition Methods in Homotopy Groups of Spheres, *by* H. TODA
50. Cohomology Operations, *lectures by* N. E. STEENROD, *written and revised by* D. B. A. EPSTEIN
51. Morse Theory, *by* J. W. MILNOR
52. Advances in Game Theory, *edited by* M. DRESHER, L. SHAPLEY, *and* A. W. TUCKER
53. Flows on Homogeneous Spaces, *by* L. AUSLANDER, L. GREEN, F. HAHN, *et al.*
54. Elementary Differential Topology, *by* J. R. MUNKRES
55. Degrees of Unsolvability, *by* G. E. SACKS
56. Knot Groups, *by* L. P. NEUWIRTH
57. Seminar on the Atiyah-Singer Index Theorem, *by* R. S. PALAIS
58. Continuous Model Theory, *by* C. C. CHANG *and* H. J. KEISLER
59. Lectures on Curves on an Algebraic Surface, *by* DAVID MUMFORD
60. Topology Seminar, Wisconsin, 1965, *edited by* R. H. BING *and* R. J. BEAN

CONTINUOUS MODEL THEORY

BY

Chen Chung Chang

AND

H. Jerome Keisler

PRINCETON, NEW JERSEY

PRINCETON UNIVERSITY PRESS

1966

L. C. CARD: 66-17698

PRINTED IN THE UNITED STATES OF AMERICA

To

Marjorie and Lois

PREFACE

This monograph contains a study of theories of models with truth values in compact Hausdorff spaces X. This collection of results might be called a theory of model theories.

It is well known to those with interests in the two-valued model theory that there has been a tremendous development in the field in the past fifteen years. While the results which led to this development range over a reasonably wide spectrum, there has been a discernible central theme among them. This theme concerns the determination of the exact relationship between the algebraic properties that a class K of models possesses and purely syntactical descriptions of sets Δ of first-order sentences by which K may be characterized. Thus many of the results are of the following type: for a class K to be closed under certain algebraic relations it is necessary and sufficient that the sentences of Δ may be restricted to those of a certain form. A basic tool needed for all of these results is the compactness theorem. More recently, a uniform method of proof for these results has been given in terms of special models.

In this monograph we have carried out a study culminating in several results along this theme for classes $\mathcal{M}$ of models which are associated with a space X of truth values. In the case that $X = \{0, 1\}$, the models in $\mathcal{M}$ will turn out to be the ordinary two-valued models. More generally, in case X is an arbitrary set, a model A in $\mathcal{M}$ is understood to be a non-empty set R and a sequence $\mathcal{A}$ of functions mapping n-tuples of R into X. As can be expected, not too much can be done if X does not carry any additional structure. If we assume, however, that X has a compact Hausdorff topology, then with continuous functions on X playing the role of connectives and with continuous set functions on X playing the role of quantifiers, we can first define a set Σ of sentences. Once we have the notions of a model A in $\mathcal{M}$ and of a sentence φ in Σ, we can proceed to introduce and study the satisfaction function which assigns a truth value $\varphi[A]$ in X to each pair φ, A. The truth value $\varphi[A]$ naturally gives rise to a function $[A]$ in X^{Σ} defined by $[A](\varphi) = \varphi[A]$. We may then consider $[\ ]$ itself as a function mapping $\mathcal{M}$ into X^{Σ}. Let X^{Σ} have the usual product topology coming from the topology on X. Then the function $[\ ]$ induces a natural topology on $\mathcal{M}$ which we call the elementary topology on $\mathcal{M}$. Using the fundamental construction of an ultraproduct of a collection of models, which is always possible when X is compact and Hausdorff, we prove that the elementary topology $\mathcal{M}$ is compact. This is the generalization of the compactness theorem in two-valued model theory, and it is one of the main tools for the rest of our work. A second important tool is our construction of special models in our theory. These results enable us to develop an extensive and fruitful theory of models with truth values in X.

It turns out that the study of models in $\mathcal{M}$ using the sentences of Σ is almost an exact analog of the study of two-valued models using sentences of the ordinary first-order predicate logic. The analogy is perfect as far as the statements of the theorems are concerned; however, the proofs are far from analogous to the corresponding two-valued proofs. In many cases the new proofs are considerably more subtle and delicate, and they give further insight into why the standard two-valued proofs work the way they do. We stress that not only do the model-theoretical theorems in this monograph look and sound like their counterparts in the two-valued model theory, but that they are indeed true generalizations of their counterparts because they will immediately specialize and yield their counterparts as consequences in the special case $X = \{0, 1\}$.

We should point out that while the classical two-valued logic has a well developed theory of expressions (i.e. syntax) in terms of the notions of deducibility, provability, and axiomatizability, in general for an arbitrary space X no such theory of expressions is known for Σ. Thus, there is an entirely separate problem of syntax for Σ which is as yet untouched in the general case.

This monograph is designed to be read and understood by anyone who has a reasonable background in point set topology and naive set theory. In particular, it can be read even without knowing the classical two-valued model theory. However, as in all branches of mathematics, it is preferable to know the classical theory first as a source of examples and motivation.

In general the titles of the chapters and sections are indicative of their contents. The first three chapters are preliminary in nature. Chapters 4 and 5 contain basic model-theoretical results leading to the deeper results of Chapters 6 and 7. Parts of the results in Chapters 5, 6, and 7 depend on the generalized continuum hypothesis. Exercises are interspersed throughout; some of them are routine while others, indicated by an asterisk, require some degree of originality. Some of the routine exercises are used in proofs later on in the text. In lieu of footnotes, a collection of historical notes can be found after Chapter 7.

Models of the classical infinite valued logic with the set $X = [0, 1]$ of truth values were first studied by Chang [1961]. In his initial study of these models, Chang used extensively some ideas of Keisler [1961], [1964], and [1964a] in two-valued model theory. Later the authors together forged the results presented herein. The authors began their joint venture at Berkeley in the Spring of 1961. The main results that they obtained at that time were announced in Chang-Keisler [1962]. Since that time their original results were considerably improved and extended. Chang gave a talk summarizing the work in the monograph at the International Symposium for Model Theory held at Berkeley, California, during June 24—July 12, 1963. A joint paper Chang-Keisler [1965] on the talk will appear in the Proceedings of the Symposium. The authors worked together again in the Summer of 1962 at Princeton and in the Summer of 1963 at Madison. During the writing of the monograph, Chang was supported in part by NSF grants and an NSF Senior Postdoctoral Fellowship for the year 1962-63 at the Institute for Advanced Study, Princeton, and Keisler was supported in part by NSF grants.

TABLE OF CONTENTS

CHAPTER VII: CLASSES PRESERVED UNDER ALGEBRAIC RELATIONS

CONTINUOUS MODEL THEORY

CHAPTER I

TOPOLOGICAL PRELIMINARIES

1.1. Notation

Throughout the monograph our mathematical notation does not differ drastically from the standard notation in general use. Thus, there is little point in giving an exhaustive encyclopedia of notation in the very first chapter. However, we shall give an idea of our notation and some highlights thereof; we hope that all existing peculiarities in our notation which do not agree with the general usage will be touched upon in the following paragraphs.

The symbols $\in$, $\cup$, $\cap$, $-$, $\bigcup$, $\bigcap$, and $\{x: \ldots x \ldots\}$ shall have their usual set-theoretical meanings. We use $\subset$ to denote the (not necessarily proper) subset relation. We shall work in naive set theory with the axiom of choice. We identify each ordinal with the set of all smaller ordinals and we define a cardinal as an initial ordinal. 0 is used for the empty set as well as the smallest ordinal. The smallest infinite ordinal is denoted by ω, and the members of ω are, in their natural order, $0, 1, 2, \ldots$. The letters m, n shall denote finite ordinals and cardinals. In general, small Greek letters $\kappa, \pi, \mu, \nu, \xi, \zeta$ denote ordinals, and the letters α, β, γ sometimes with subscripts, are reserved for cardinals. Given ordinals μ, ν and cardinals α, β, $\mu < \nu$ shall mean that μ is smaller than ν, $\mu + \nu$ shall denote the ordinal sum of μ and ν, α^+ the cardinal successor of α, α^β the cardinal α to the power β, $\mathrm{cf}(\alpha)$ the cofinality of α, and $\Sigma_{\beta<\alpha} 2^\beta$ the cardinal sum of the cardinals 2^β for each $\beta < \alpha$. Since the cardinal $\Sigma_{\beta<\alpha} 2^\beta$ depends only on α, it is also denoted by α^*. The cardinal sum, or maximum, of two infinite cardinals α, β is given by their union $\alpha \cup \beta$.

Let X be a set and α a cardinal. The cardinality, or power, of X shall be denoted by $|X|$. We let

$$S(X) = \{Y: \ Y \subset X\} \quad ,$$

$$X^* = S(X) - \{0\} \quad ,$$

$$S_\alpha(X) = \{Y: \ Y \subset X \text{ and } |Y| < \alpha\} \quad ,$$

$$S^\alpha(X) = \{Y: \ Y \subset X \text{ and } |X-Y| < \alpha\} \quad .$$

Binary relations over a set X are defined as sets of ordered pairs $\langle x, y \rangle$ of elements x, y belonging to X. The domain and range of a (binary) relation H will be denoted by $\mathscr{D}H$ and $\mathscr{R}H$, respectively. The converse of a relation H is denoted by $\breve{H}$ and the relative product, or composition, of two relations G and H by $G \circ H$; see, for example, Kelley [1955]. If Y is a set and H is a relation, we let

$$H[Y] = \{y: \langle x, y \rangle \in H \text{ for some } x \in Y\} \quad .$$

It will be convenient to write $H[x]$ for $H[\{x\}]$ when $x \in \mathscr{D}H$ and no confusion can result. We sometimes write $x\,H\,y$ for $\langle x, y \rangle \in H$.

Functions are defined as special binary relations in the usual manner. Thus, the notations $\mathscr{D}f$, $\mathscr{R}f$, $\breve{f}$, $f \circ g$, and $f[X]$ may be applied to functions f, g. A function f with domain X and range included in Y is sometimes graphically presented as $f: X \to Y$. The expression $(\lambda x \in X)(\ldots x \ldots)$ shall denote the function with domain X which maps each $x \in X$ into $(\ldots x \ldots)$. If f is a function with domain X and Y is any set, we define $f \restriction Y = (\lambda y \in X \cap Y)f(y)$. The Cartesian product of sets Y_i with $i \in I$ is denoted by $\Pi_{i \in I} Y_i$, and the Cartesian power of X to the Y by X^Y. Thus the expressions $f \in X^Y$ and $f: Y \to X$ have the same meanings and may be used interchangeably. For a finite Cartesian power, we identify X^n with $X \times X \times \ldots \times X$ n-times, so the elements of X^n may be considered as ordered n-tuples. In particular, if $n = 1$, then we identify X^1 with X and $\langle x \rangle$ with x. The special notation X^∞ is used to denote the set of functions $f \in X^\omega$ such that f is eventually constant. That is,

$$X^{\infty} = \{f \in X^{\omega}: \text{ for some } n, \ f(n) = f(m) \text{ for all } m \geq n\} \quad .$$

An *enumeration* of a set X is a function whose domain is an ordinal and whose range is X.

1.2. D-products

D is said to be a *filter* over the set I if:

(i) $D \subset S(I)$ and $0 \neq D \neq S(I)$;

(ii) whenever $X \in D$ and $X \subset Y$, then $Y \in D$;

(iii) whenever $X, Y \in D$, then $X \cap Y \in D$.

In the literature the condition $D \neq S(I)$ is often dropped from the definition of filter, and filters in the above sense are called proper filters. A filter D over I is called *maximal* if for every $X \subset I$, either $X \in D$ or $I - X \in D$. Maximal filters are also known as *ultrafilters*. A filter D is *principal* if $\bigcap D \in D$. We assume the reader is familiar with the elementary properties of ultrafilters. For instance, he should know that ultrafilters exist over any non-empty set I. In fact, any subset E of S(I) which has the finite intersection property (that is, no finite intersection of elements of E is empty) can be extended to an ultrafilter. A filter D over the set $S_{\omega}(I)$ is said to be *regular* if for every $j \in I$, we have

$$\{i \in S_{\omega}(I): j \in i\} \in D \quad .$$

The phrase "regular filter over J" is meaningful only when J is a set of the form $S_{\omega}(I)$. Quite obviously, regular filters exist over any set $S_{\omega}(I)$ because the family of sets

$$\{\{i \in S_{\omega}(I) : j \in i\} : j \in I\}$$

has the finite intersection property.

In the following series of exercises we mention some other special kinds of ultrafilters which have arisen in the literature and which, like regular ultrafilters, are of interest when D-products are studied.

EXERCISE 1A. If I is finite, then the only regular ultrafilter over $S_{\omega}(I)$ is the principal filter generated by $\{I\}$.

EXERCISE 1B*. Let us call an ultrafilter D over I *weakly regular* if there is a function $f: I \to S_{\omega}(I)$ such that for all $j \in I$,

$$\check{f}[\{i \in S_\omega(I) : j \in i\}] \in D \quad .$$

(a) Every regular ultrafilter over a set $S_\omega(J)$ is weakly regular.

(b) An ultrafilter D over a set I is weakly regular iff there exists $E \subset D$ such that $|E| = |I|$ and the intersection of any infinite subset of E is empty.

EXERCISE 1C. (a) Every ultrafilter over a finite set is weakly regular.

(b) An ultrafilter D over a countable set I is weakly regular iff it is non-principal.

EXERCISE 1D. An ultrafilter D over a set I is said to be *uniform* if $|X| = |I|$ for all $X \in D$. D is uniform iff $S^{|I|}(I) \subset D$. A finite set with more than one element has no uniform ultrafilters over it. However, every weakly regular ultrafilter over an infinite set I is uniform.

EXERCISE 1E. Let $I = \bigcup_{j \in J} I_j$, let E be an ultrafilter over J, and let D_j be an ultrafilter over I_j for each $j \in J$. Then the set

$$D = \{X \subset I : \{j \in J : X \cap I_j \in D_j\} \in E\}$$

is an ultrafilter over I.

We shall now give the definition of a D-product. Let I be a non-empty set, D a filter over I, and R_i a non-empty set for each $i \in I$. Let $R' = \Pi_{i \in I} R_i$. For $f, g \in R'$, we say that $f \sim g$ (read f is equivalent to g) if $\{i: f(i) = g(i)\} \in D$. We remark that the symbol $\sim$ is always used relative to some filter D given by context. It is easy to verify that $\sim$ is an equivalence relation over R'. For $f \in R'$, let $f^\sim = \{g \in R': f \sim g\}$. We then define the D-*product* of the function $(\lambda i \in I)R_i$ by

$$\text{D-prod } \lambda i R_i = \{f^\sim : f \in R'\} \quad .$$

Since I is determined by D, there is no confusion in writing λi instead of $\lambda i \in I$ in our above notation. If each R_i coincides with the same set R, the D-product of $(\lambda i \in I)R$ is called the D-*power* of R and is denoted by D-prod R.

In the literature, D-products are called reduced direct products, or reduced products, and D-powers are called reduced (direct) powers. In case D is an ultrafilter, D-products are called ultraproducts, or prime (reduced direct) products, and D-powers are called ultrapowers, or prime (reduced direct) powers. There is a large collection of theorems, and even some open problems, concerning the cardinality of D-products. We shall prove only two very weak results which we shall need later on. However, a number of other results in this direction will be indicated in the series of exercises at the end of this section.

LEMMA 1.2.1. For any filter D over a set I and any non-empty set R, $|R| \leq |\text{D-prod } R|$.

PROOF. The function $f = (\lambda r \in R)((\lambda i \in I)r)^{\sim}$, which maps each $r \in R$ into the equivalence class of the constant function with range $\{r\}$, is a one-to-one function on R into D-prod R.

LEMMA 1.2.2. Suppose that R is an infinite set, $S_\omega(J) = I$, and D is a regular ultrafilter over I. Then $|\text{D-prod } R| \geq |I|$.

PROOF. We may assume that $|I| \geq \omega$, so that $|I| = |J|$. We may also assume that R contains an infinite sequence

$$a_0, a_1, a_2, \ldots, a_m, \ldots$$

of distinct elements, and that J has been simply ordered in some fashion. For each $j \in J$ and $i \in I$, we define

$$f_j(i) = \begin{cases} a_0 & \text{if } j \notin i, \\ a_m & \text{if } j \text{ is the m-th element in the ordering induced on } i. \end{cases}$$

Thus, for each $j \in J$, $f_j \in R^I$. Let $b_j = (f_j)^{\sim}$. Suppose that $j_1 \neq j_2$. Since D is regular, the set

$$I' = \{i \in I : j_1, j_2 \in i\} \in D .$$

Now, whenever $i \in I'$, we have $f_{j_1}(i) \neq f_{j_2}(i)$. Hence

$$\{i: f_{j_1}(i) \neq f_{j_2}(i)\} \in D ,$$

and $b_{j_1} \neq b_{j_2}$. This proves the lemma.

EXERCISE 1F. Suppose D is an ultrafilter over I. If either R or I is finite, then $|D\text{-prod } R| = |R|$. If D is principal, then $|D\text{-prod } R| = |R|$. If D is any ultrafilter over I, then $|D\text{-prod } R| \leq |R^I|$.

EXERCISE 1G*. If D is a uniform ultrafilter over an infinite set I, then $|D\text{-prod } I| > |I|$. Hint: given a sequence of $|I|$ functions $f: I \to I$, construct a new function $g: I \to I$ which is not equivalent to any of them.

EXERCISE 1H*. Under the same hypotheses as in 1.2.2, prove that $|D\text{-prod } R| > |I|$. Indeed one can prove the following stronger result: If R is an infinite set and D is a weakly regular ultrafilter over a set J, then $|D\text{-prod } R| = |R^J|$. Hint: let $f: J \to S_\omega(J)$ be as in Ex. 1B; associate with each $g \in R^J$ a function $g' \in R^J$ in such a way that, whenever $g(j) \neq h(j)$ and $j \in f(i)$, we have $g'(i) \neq h'(i)$.

EXERCISE 1I*. If $|R| = |R|^\alpha$, then $|D\text{-prod } R| = |D\text{-prod } R|^\alpha$. Hint: a function f on R onto R^α induces in a natural way a function f' on R^I onto $(R^I)^\alpha$, and f' in turn induces a function f'' on D-prod R onto $(D\text{-prod } R)^\alpha$.

EXERCISE 1J. An ultrafilter D is said to be *countably incomplete* if there exists a countable subset $E \subset D$ such that $\cap E \notin D$. Any weakly regular filter on an infinite set I is countably incomplete. No principal ultrafilter is countably incomplete.

EXERCISE 1K*. If D is a countably incomplete ultrafilter over a set I, and R is an infinite set, then $|D\text{-prod } R| = |D\text{-prod } R|^\omega$. If E is a countably complete ultrafilter and S is countable, then E-prod S is countable. Hint: show that there is a countably decreasing sequence of members of D whose intersection is empty, and then argue as in Exercise 1I.

1.3. Compact Hausdorff spaces

We assume that the reader is familiar with all of the elementary parts of Kelley [1955], particularly with chapters one through five. We shall follow the topological notation of that book, and where we differ from it or where we introduce new notions, we shall so indicate at the appropriate places.

Let X be a compact Hausdorff topological space with topology $\mathcal{S}$. Let $\mathcal{S}_0$ be an open basis for $\mathcal{S}$; for convenience, we assume $X \in \mathcal{S}_0$. We shall identify X with the pair $(X, \mathcal{S})$ and assume that X and $\mathcal{S}_0$ are held fixed throughout the monograph. Thus, $\mathcal{S}$ is the collection of all open sets of X; the letters U, V, W, sometimes with subscripts, shall denote open sets. Elements of $\mathcal{S}_0$ are referred to as *basic open sets* and complements of elements of $\mathcal{S}_0$ are called *basic closed sets*. The closure of a subset Y of X is denoted by $\bar{Y}$. For each n let X^n denote the n-fold product space of X with the usual product topology. Each X^n is compact and Hausdorff. For arbitrary topological spaces Y and Z, let $\mathcal{C}(Y, Z)$ be the set of all continuous functions $f\colon Y \to Z$. If Y is compact and Z is Hausdorff, then every member of $\mathcal{C}(Y, Z)$ is closed in the sense that the image of each closed set of Y is a closed set of Z. For each n let $\mathcal{C}_n = \mathcal{C}(X^n, X)$, and let $\mathcal{C} = \bigcup_{n<\omega} \mathcal{C}_n$; $\mathcal{C}$ is the set of all continuous functions of finitely many variables ranging over X, and with values in X.

Every compact Hausdorff space Y is *normal*, that is any two disjoint closed sets in Y can be extended to two disjoint open sets. Urysohn's Lemma states that *if* Y *is a normal space and* U, V *are disjoint closed sets in* Y, *then there is a continuous function* $f\colon Y \to [0, 1]$ *such that* $f[U] = \{0\}$ *and* $f[V] = \{1\}$. (See Kelley [1955], pages 115 and 141; here $[0, 1]$ denotes the unit interval on the real line.)

Given a subset $\mathcal{F} \subset \mathcal{C}$, the *closure of* $\mathcal{F}$ *under composition* is defined as the least set $\mathcal{G} \supset \mathcal{F}$ such that:

1) Whenever $m \leq n$, the m-th projection function $(\lambda \langle x_1, \ldots, x_n\rangle \in X^n)(x_m)$ on X^n onto X belongs to $\mathcal{G}$;

2) if $f_1, \ldots, f_n \in \mathcal{G} \cap \mathcal{C}_m$ and $g \in \mathcal{F} \cap \mathcal{C}_n$, then the function $(\lambda z \in X^m) g(f_1(z), \ldots, f_n(z))$ belongs to $\mathcal{G}$.

We always have $\mathcal{G} \subset \mathcal{C}$, that is, a composition of continuous functions is continuous. Moreover, the closure of $\mathcal{G}$ under composition is $\mathcal{G}$ itself.

We let $\mathcal{S}^*$ be the least topology on the set X^* such that:

1) Whenever V is open in X, the set $\{Y \in X^*: \bar{Y} \subset V\}$ is open in $\mathcal{S}^*$; and

2) whenever U is open in X, the set $\{Y \in X^*: Y \cap U \neq 0\}$ is open in $\mathcal{S}^*$.

When we speak of the topological space X^*, we mean the space $(X^*, \mathcal{S}^*)$. An open basis for the space X^* is the family $\mathcal{S}_0^*$ of all sets of the form

$$\{Y \in X^* : \bar{Y} \subset V_1 \cup \ldots \cup V_n \text{ and } Y \cap V_m \neq 0 \text{ for each } m \leq n\}$$

where $V_1, \ldots, V_n \in \mathcal{S}_0$. The space X^* is in general not Hausdorff. For example, if $Y \in X^*$, then Y and $\bar{Y}$ cannot be separated by disjoint open neighborhoods; indeed, Y belongs to an open set of X^* if and only if $\bar{Y}$ does. On the other hand, if Y and Z are distinct closed non-empty sets of X, then they do have disjoint open neighborhoods. Thus if we restrict our attention to the closed sets $Y \in X^*$, we have a Hausdorff space. As we shall prove in Section 1.5, the space X^* is compact. Since X is Hausdorff, every member f of $\mathcal{C}(X^*, X)$ is closed and, furthermore, $f(Y) = f(\bar{Y})$ for all $Y \in X^*$. We define $\mathcal{Q} = \mathcal{C}(X^*, X)$.

We shall now prove an easy limit theorem for the space X^*.

THEOREM 1.3.1. Let $Y \in X^{**}$ have the property that, if $Z_1, Z_2 \in Y$, then there is a set $Z_3 \in Y$ such that $Z_1 \cup Z_2 \subset Z_3$. Then $\cup Y \in \bar{Y}$, where $\bar{Y}$ is the closure of Y in the space X^*.

PROOF. Let $\cup Y \in U \in \mathcal{S}_0^*$; we must show that $U \cap Y \neq 0$. Choose basic open sets $V_1, \ldots, V_n \in \mathcal{S}_0$ such that

$$U = \{Z \in X^* : \bar{Z} \subset V_1 \cup \ldots \cup V_n \text{ and } Z \cap V_m \neq 0 \text{ for } m \leq n\} .$$

Let $y_m \in \cup Y \cap V_m$ for $m \leq n$. Choose sets $Z_1, \ldots, Z_n \in Y$ such that $y_m \in Z_m$ for each $m \leq n$. By hypothesis, there is a set $Z \in Y$ with $Z_1 \cup \ldots \cup Z_n \subset Z$. Then we have $Z \cap V_m \neq 0$ for $m \leq n$, and

$$\bar{Z} \subset \overline{\cup Y} \subset V_1 \cup \ldots \cup V_n .$$

Hence $Z \in U \cap Y$, and our proof is complete.

Let us now consider, for any non-empty set I, the product space X^I with the usual product topology, which we shall denote by $\mathcal{S}^I$. By Tychonoff's theorem, the space $(X^I, \mathcal{S}^I)$ is compact and Hausdorff. An open set V in X^I will be called *singular* if, for some $i \in I$ and $U \in \mathcal{S}_0$,

$$V = \{h \in X^I : h(i) \in U\} \quad .$$

An open set V in X^I is called *basic* if it is a finite intersection of singular open sets. The set of basic open sets in X^I is the usual open base for the space X^I, and the set of singular open sets is the usual open subbase for X^I. A closed set Y in X^I is called *singular*, or *basic*, if its complement is a singular, or basic, open set, respectively. Each closed set of X^I is then an intersection of basic closed sets. It is easily seen that the power of the set of basic open sets (or basic closed sets) in the space X^I is at most $\omega \cup |\mathcal{S}_0| \cup |I|$.

1.4. Ordered spaces

Let H be a binary relation over the set X, and let $\underset{\sim}{0}, \underset{\sim}{1} \in X$. We shall say that the triple $(X, \underset{\sim}{0}, \underset{\sim}{1})$ is *ordered by* H if the following conditions are satisfied:

1) H is reflexive and transitive over X ;

2) H is *open*, that is, whenever $V \in \mathcal{S}$, then $H[V] \in \mathcal{S}$;

2') $\breve{H}$ is open ;

3) $\underset{\sim}{1}$ is a *continuous fixed point* of H, that is, whenever $\underset{\sim}{1} \in V \in \mathcal{S}$, there is an open set $U \in \mathcal{S}$ such that $\underset{\sim}{1} \in H[U] \subset V$;

3') $\underset{\sim}{0}$ is a continuous fixed point of $\breve{H}$, that is, whenever $\underset{\sim}{0} \in V \in \mathcal{S}$, there is a set $U \in \mathcal{S}$ such that $\underset{\sim}{0} \in \breve{H}[U] \subset V$.

Notice that the triple $(X, \underset{\sim}{0}, \underset{\sim}{1})$ is always ordered by the identity relation over X.

EXERCISE 1L. Assume 1) and 2) above. $\underset{\sim}{1}$ is a continuous fixed point of H iff $H[\underset{\sim}{1}] = \{\underset{\sim}{1}\}$ and for every function $F \subset H$, $F\colon X \to X$, F is continuous at the point $\underset{\sim}{1}$. This justifies the terminology introduced in 3) and 3') above.

A function $f \in \mathcal{C}_n$ is said to be H-*preserving* if the ordered pair $< f(x_1, \ldots, x_n), f(y_1, \ldots, y_n) >$ belongs to H whenever $< x_1, y_1 >, \ldots, < x_n, y_n > \in H$. We let $\mathcal{C}_H$ denote the set of all H-preserving functions in $\mathcal{C}$. A function $f \in \mathcal{Q}$ is said to be H-*preserving* if we have $f(Y)Hf(Z)$ whenever $Y, Z \in X^*$ and $Z \subset H[Y]$. We let $\mathcal{Q}_H$ be the set of all H-preserving functions in $\mathcal{Q}$.

We may extend the relation H over X to a relation H^* over the set X^* as follows:

$$H^* = \{< Y, Z > : Y, Z \in X^*, Z \subset H[Y], \text{ and } Y \subset \breve{H}[Z]\} \quad .$$

The following properties of the $*$ operation on relations are immediate. xHy if and only if $\{x\}H^*\{y\}$. If H is reflexive, then so is H^*. If H is symmetric, then so is H^*. If H is transitive, then so is H^*. In general, we have the equation $(\breve{H})^* = (H^*)\breve{}$. A function $f \in \mathcal{Q}$ is said to be H^*-*preserving* if $f(Y)\ Hf(Z)$ whenever $Y\ H^*\ Z$. We let $\mathcal{Q}_{H^*}$ be the set of all H^*-preserving functions in $\mathcal{Q}$. It is clear that every H-preserving function in $\mathcal{Q}$ is H^*-preserving, that is $\mathcal{Q}_H \subset \mathcal{Q}_{H^*}$. Moreover, it is easily seen that $\mathcal{Q}_{H^*} = \mathcal{Q}_{\breve{H}^*}$, so that $\mathcal{Q}_{\breve{H}} \subset \mathcal{Q}_{H^*}$.

EXERCISE 1M. If the relation H is a function, then H^* is the function $(\lambda Y \in X^*)H[Y]$.

We conclude this section with a lemma which we shall need in Chapter 7, and an exercise.

LEMMA 1.4.1. Let $\underset{\sim}{0}, \underset{\sim}{1} \in X$, and assume that

(i) $(X, \underset{\sim}{0}, \underset{\sim}{1})$ is ordered by H ;

(ii) whenever $< x, y > \in X^2 - H$, there is an H-preserving function $f \in \mathcal{C}_1 \cap \mathcal{C}_H$ such that $f(x) = \underset{\sim}{1}$ and $f(y) \neq \underset{\sim}{1}$.

Then the relations H and $\breve{H}$ are closed, that is, if Y is closed in X, then $H[Y]$ and $\breve{H}[Y]$ are closed in X.

PROOF. Let Y be a closed set. We first prove that $H[Y]$ is closed. Let $z \notin H[Y]$. It is sufficient to find a $U \in \mathcal{S}$ such that $z \in U$ and $U \cap H[Y] = 0$. Let $y \in Y$. Since $z \notin H[Y]$, $< y, z > \notin H$. There is a function $f \in \mathcal{C}_1 \cap \mathcal{C}_H$ such that $f(y) = \underset{\sim}{1}$ and $f(z) \neq \underset{\sim}{1}$. Pick V_y and U_y in $\mathcal{S}$ so that $\underset{\sim}{1} \in V_y$, $f(z) \in U_y$, and $V_y \cap U_y = 0$. By hypothesis, we may as well assume that $H[V_y] \cap U_y = 0$. Hence, $f^{\cup}[H[V_y]] \cap f^{\cup}[U_y] = 0$. Since f is H-preserving, $H[f^{\cup}[V_y]] \subset f^{\cup}[H[V_y]]$. Let $V'_y = f^{\cup}[V_y]$ and $U'_y = f^{\cup}[U_y]$. We see that $y \in V'_y$, $z \in U'_y$, V'_y and U'_y are open and disjoint, and $H[V'_y] \cap U'_y = 0$. Now $\{V'_y : y \in Y\}$ is an open covering of Y. Hence, there are $y_1, \ldots, y_n \in Y$ such that

$$Y \subset V'_{y_1} \cup \ldots \cup V'_{y_n} \quad .$$

Thus,

$$H[Y] \subset H[V'_{y_1}] \cup \ldots \cup H[V'_{y_n}] \quad .$$

Let $U = U'_{y_1} \cap \ldots \cap U'_{y_n}$. Clearly $z \in U \in \mathcal{S}$ and $U \cap H[Y] = 0$. To prove that $\breve{H}[Y]$ is closed, we argue in an analogous manner and use the following simple fact. For any two sets U and V, $H[U] \cap V = 0$ if and only if $U \cap \breve{H}[V] = 0$. The lemma is proved.

We remark that in the proof of the above lemma, it is sufficient to assume that $\underset{\sim}{1}$ is a continuous fixed point of H and condition (ii).

EXERCISE 1N. Let the triple $(X, \underset{\sim}{0}, \underset{\sim}{1})$ be ordered by H. Let I be a non-empty set, define the relation H pointwise over the product space X^I, and let $\underset{\sim}{0}'$, $\underset{\sim}{1}'$ be the constant functions in X^I with values $\underset{\sim}{0}$, $\underset{\sim}{1}$.

(a) The triple $(X^I, \underset{\sim}{0}', \underset{\sim}{1}')$ is ordered by H.

(b) If $(X, \underset{\sim}{0}, \underset{\sim}{1})$ satisfies the hypothesis (ii) of Lemma 1.4.1, then so does $(X^I, \underset{\sim}{0}', \underset{\sim}{1}')$, and hence H and $\breve{H}$ are closed over X^I.

1.5. D-limits

Let I be an arbitrary non-empty index set and let D be an ultrafilter over I. For each function $f \in X^I$, we shall define the D-limit of f as a certain point of X. We shall extend the operation of taking D-limits to functions $F \in X^{*I}$; the extended operation is called the D*-limit. This section is devoted to some elementary properties of D-limits and D*-limits.

THEOREM 1.5.1. For each function $f \in X^I$ there corresponds exactly one point $x \in X$ with the following property:

(i) for each neighborhood V of x, $\{i: f(i) \in V\} \in D$.

PROOF. First of all, let us show that there exists at most one point of X having the property (i). Suppose that x and y are two distinct points of X both having property (i). Then we may choose disjoint neighborhoods V_x of x and V_y of y. Now,

$$\{i: f(i) \in V_x\} \in D \text{ and } \{i: f(i) \in V_y\} \in D \quad .$$

Hence

$$\{i: f(i) \in V_x\} \cap \{i: f(i) \in V_y\} = \{i: f(i) \in V_x \cap V_y\} = 0 \in D \quad .$$

This is a contradiction.

To prove existence let us assume that, to the contrary,

(1) each $x \in X$ has a neighborhood V_x such that

$$\{i: f(i) \in V_x\} \notin D \, .$$

The family $\{V_x: x \in X\}$ is clearly an open covering of X. Therefore, for some finite number of elements $x_1, \ldots, x_n \in X$ we have

(2) $V_{x_1} \cup \ldots \cup V_{x_n} = X$.

Since $\mathcal{R}f \subset X$, (2) implies that

$$\{i: f(i) \in V_{x_1}\} \cup \ldots \cup \{i: f(i) \in V_{x_n}\} = I \quad .$$

Since D is an ultrafilter, there exists an $m \leq n$ such that $\{i: f(i) \in V_{x_m}\} \in D$. Thus (1) cannot hold and the theorem is proved.

For each $f \in X^I$, we define the *D-limit of* f, written D-lim f, as the unique point $x \in X$ satisfying property (i) of Theorem 1.5.1. Since I is determined uniquely by D, we sometimes write D-lim $\lambda i\, f(i)$ for D-lim f. For a set of functions $Y \subset X^I$, we put D-lim Y = {D-lim f: $f \in Y$}.

LEMMA 1.5.2. For each $f \in X^I$, D-lim $f \in \overline{\mathcal{R}f}$. In particular, the D-limit of a constant function is the value of the function.

PROOF. It is clear from the definition that D-lim f is either in $\mathfrak{R}f$ or is an accumulation point of $\mathfrak{R}f$.

THEOREM 1.5.3. Let $g \in \mathcal{C}_n$ and let $f_1, \ldots, f_n \in X^I$. Then

$$\text{D-lim } \lambda i\; g(f_1(i), \ldots, f_n(i)) = g(\text{D-lim } f_1, \ldots, \text{D-lim } f_n) \quad .$$

PROOF. It is sufficient to prove that, for every neighborhood V of the point $g(\text{D-lim } f_1, \ldots, \text{D-lim } f_n)$, we have

(1) $$\{i\colon g(f_1(i), \ldots, f_n(i)) \in V\} \in D \quad .$$

Since $g \in \mathcal{C}_n$, there exist neighborhoods U_m of D-lim f_m, $m = 1, \ldots, n$, such that $g(y) \in V$ whenever $y \in U_1 \times \ldots \times U_n$. Also, for these $U_1, \ldots, U_n$, we have

$$I_m = \{i\colon f_m(i) \in U_m\} \in D \text{ for each } m \leq n \quad ,$$

hence $I_1 \cap \ldots \cap I_n \in D$. If $i \in I_1 \cap \ldots \cap I_n$, then $g(f_1(i), \ldots, f_n(i)) \in V$. Thus (1) holds, and the theorem is proved.

Now we turn to the operation D-lim Y.

LEMMA 1.5.4. Let $F \in X^{*I}$ and let $Y = \text{D-lim } \Pi_{i\in I} F(i)$. Then $\text{D-lim } \Pi_{i\in I} \overline{F(i)} \subset \bar{Y}$.

PROOF. Suppose $f \in \Pi_{i\in I} \overline{F(i)}$, and let V be a neighborhood of the point D-lim f. We must show that $Y \cap V \neq 0$. Since X is a normal space and the sets {D-lim f}, $X - V$ are closed, there exists an open neighborhood U of D-lim f such that $\bar{U} \subset V$. We have $\{i\colon f(i) \in U\} \in D$, and therefore we may choose a function $g \in \Pi_{i\in I} F(i)$ such that $\{i\colon g(i) \in U\} \in D$. Then D-lim $g \in Y$, and also D-lim $g \in \bar{U}$. Consequently, D-lim $g \in Y \cap V$.

THEOREM 1.5.5. Let $F \in X^{*I}$ and let $Y = \text{D-lim } \Pi_{i\in I} F(i)$. Then Y has the following property:

(i) Whenever $Y \in V \in \mathcal{S}_0^{\,*}$, $\{i\colon F(i) \in V\} \in D$.

PROOF. In view of the definition of $\mathcal{S}_0^{\,*}$, it suffices to prove the following:

(1) Whenever $U \in \mathcal{S}_0$ and $Y \cap U \neq 0$, $\{i\colon F(i) \cap U \neq 0\} \in D$;

(2) Whenever $\overline{Y} \subset V_1 \cup \ldots \cup V_n$ and $V_1, \ldots, V_n \in \mathscr{S}_0$,

$$\{i: \overline{F(i)} \subset V_1 \cup \ldots \cup V_n\} \in D \quad .$$

If $x \in Y \cap U$, then $x = D\text{-lim } g$ for some $g \in \Pi_{i\in I} F(i)$, and $\{i: g(i) \in U\} \in D$. Hence $\{i: F(i) \cap U \neq 0\} \in D$, and (1) is verified.

To prove (2), we assume that

$$J = \{i: \overline{F(i)} \subset V_1 \cup \ldots \cup V_n\} \notin D$$

and arrive at a contradiction. Choose $f \in \Pi_{i\in I} \overline{F(i)}$ such that, for each $j \in I-J$, we have $f(j) \notin V_1 \cup \ldots \cup V_n$. Then $D\text{-lim } f \notin V_1 \cup \ldots \cup V_n$. But by Lemma 1.5.4, $D\text{-lim } f \in \bar{Y}$, contradicting our assumption that $\bar{Y} \subset V_1 \cup \ldots \cup V_n$. This proves (2) and completes the proof of the theorem.

Notice that Theorem 1.5.5 cannot be formulated in the exact manner of Theorem 1.5.1. This is because the space X^* is not Hausdorff. From our discussions so far one could formulate and prove the following statement which is an exact analog of Theorem 1.5.1:

For each function $F \in X^{*I}$ there corresponds exactly one closed set Y of X with the following property:

$$\text{whenever } Y \in V \in \mathscr{S}_0^*, \quad \{i: F(i) \in V\} \in D \quad .$$

For each function $F \in X^{*I}$, we define the D^*-*limit of* F, in symbols $D^*\text{-lim } F$, as the set $D\text{-lim } \Pi_{i\in I} F(i)$.

THEOREM 1.5.6. Let $q \in \mathfrak{Q}$ and $F \in X^{*I}$. Then

$$D\text{-lim } \lambda i\, q(F(i)) = q(D^*\text{-lim } F) \quad .$$

PROOF. It is sufficient to prove that for every neighborhood V of $q(D^*\text{-lim } F)$ we have $\{i: q(F(i)) \in V\} \in D$. Since $q \in \mathfrak{Q}$, there exists a $U \in \mathscr{S}_0^*$ such that $D^*\text{-lim } F \in U$ and $q[U] \subset V$. By Theorem 1.5.5, $\{i: F(i) \in U\} \in D$. Therefore $\{i: q(F(i)) \in V\} \in D$. This proves the theorem.

All of our results so far hold for an arbitrary non-empty index set I, and an arbitrary ultrafilter D over I. In order to derive some other results, we have to select the set I and the ultrafilter D with some discrimination. Of particular interest is the case when $I = S_\omega(J)$ for some non-empty set J and D is a regular ultrafilter over I.

LEMMA 1.5.7. Suppose that

(i) $I = S_\omega(J)$ where J is non-empty ;

(ii) D is a regular ultrafilter over I ; and

(iii) $F \in X^{*I}$ is such that $F(i \cup i') \subset F(i) \cap F(i')$ for all $i, i' \in I$.

Then $D^*\text{-lim } F \subset \bigcap_{j \in J} \overline{F(\{j\})}$.

PROOF. Let $f \in \Pi_{i \in I} F(i)$, let V be any neighborhood of D-lim f, and let $j \in J$. It suffices to prove that $F(\{j\}) \cap V \neq 0$. First of all, we have

(1) $$\{i: f(i) \in V\} \in D \quad .$$

Next, for each $i \in I$, if $j \in i$, then $F(i) \subset F(\{j\})$ and $f(i) \in F(\{j\})$. Hence

(2) $$\{i: j \in i \text{ and } i \in I\} \subset \{i: f(i) \in F(\{j\})\} \quad .$$

Since D is regular, $\{i: j \in i \text{ and } i \in I\} \in D$, and hence by (2)

(3) $$\{i: f(i) \in F(\{j\})\} \in D \quad .$$

(1) and (3) imply that $\{i: f(i) \in F(\{j\}) \cap V\} \in D$ and, in particular, $F(\{j\}) \cap V \neq 0$. The lemma is proved.

By examining the proof of Lemma 1.5.7, we see that it applies not only to the space X, but to the space X^* as well. We simply state the corresponding lemma without proof.

LEMMA 1.5.7*. Suppose that

(i) $I = S_\omega(J)$ where J is non-empty;

(ii) D is a regular ultrafilter over I; and

(iii) $F \in (X^*)^{*I}$ is such that $F(i \cup i') \subset F(i) \cap F(i')$ for all $i, i' \in I$.

Then $\{D^*\text{-lim } f: f \in \Pi_{i \in I} F(i)\} \subset \bigcap_{j \in J} \overline{F(\{j\})}$.

THEOREM 1.5.8. The space X^* is compact

PROOF. Let G be a non-empty family of closed sets of X^* having the finite intersection property. Let $I = S_\omega(G)$ and let D be a regular ultrafilter over I. Let $F \in (X^*)^{*I}$ be defined by the equation $F(i) = X^* \cap \bigcap i$ for each $i \in I$, and let $f \in \Pi_{i \in I} F(i)$. Clearly all hypotheses of Lemma 1.5.7* are satisfied. Therefore, by 1.5.7*,

$$D^*\text{-}\lim f \in \bigcap_{Z\in G} \overline{F(\{Z\})} = \bigcap G.$$

Thus $\bigcap G \neq 0$, and X^* is compact.

THEOREM 1.5.9. Let $I = S_\omega(J)$ and $|J| \geq |\mathscr{S}_0|$, and let D be a regular ultrafilter over I. Let $Y \subset X$ and x be an accumulation point of Y. Then there exists a function $f \in Y^I$ such that $D\text{-}\lim f = x$. Hence $D^*\text{-}\lim Y^I$ is closed in X.

PROOF. We may assume, without loss of generality, that $\mathscr{S}_0 \subset J$. Let the function $F \in X^{*I}$ be defined by the equation $F(i) = Y \cap \bigcap\{V \in \mathscr{S}_0 : x \in V \in i\}$ for each $i \in I$, with the convention $\bigcap 0 = X$, and let $f \in \Pi_{i\in I} F(i)$. By Lemma 1.5.7, for each neighborhood U of x and V of $D\text{-}\lim f$, we have $F(\{U\}) \cap V \neq 0$. That is, $Y \cap U \cap V \neq 0$. Since X is Hausdorff, this implies $D\text{-}\lim f = x$.

We shall say that a set $Y \subset X$ is *closed under* D-*limits* if, for every set I and every ultrafilter D over I, we have $D\text{-}\lim Y^I = Y$.

THEOREM 1.5.10. Let $Y \subset X$, $I_0 = S_\omega(\mathscr{S}_0)$, and D_0 be a regular ultrafilter over I_0. Then the following three conditions are equivalent:

(i) Y is a closed set of X ;

(ii) Y is closed under D-limits;

(iii) $D_0\text{-}\lim Y^{I_0} = Y$.

PROOF. The implication (i) to (ii) follows from Lemma 1.5.2. The implication (ii) to (iii) is trivial, and (iii) to (i) follows from Theorem 1.5.9.

CHAPTER II

CONTINUOUS LOGICS

2.1. Definition of a continuous logic

In this chapter we shall begin our study of continuous (first-order predicate) logics. We shall be concerned now with syntactical notions, and especially with the notion of a formula. In what follows, the lower case Greek letters φ, ψ, θ will be reserved for formulas, and capital Greek letters will be used for sets of formulas.

We begin by introducing the notion of a similarity type. By a *(similarity) type* we shall mean an ordered pair $< \tau, \kappa >$, where κ is an ordinal and τ is a sequence of natural numbers, i.e., a function $\tau \in \omega^{\pi}$ where π is an ordinal.

By a *continuous (first-order predicate) logic* we shall mean a sextuple

$$\mathscr{L} = (< \tau, \kappa >, X, \mathscr{F}, \underset{\sim}{0}, \underset{\sim}{1}, H) \quad ,$$

where $< \tau, \kappa >$ is a similarity type, X is a compact Hausdorff space, $\mathscr{F} \subset \mathscr{C} \cup \mathscr{Q}$, $\mathscr{F} \cap \mathscr{C}$ is closed under composition, $\underset{\sim}{0}, \underset{\sim}{1}$ are two distinct points of X, and $(X, \underset{\sim}{0}, \underset{\sim}{1})$ is ordered by H (see Section 1.4). We shall refer to $< \tau, \kappa >$ as the *type of the logic* $\mathscr{L}$, to X as the *value space of* $\mathscr{L}$, $\underset{\sim}{0}, \underset{\sim}{1}$ as the *designated values of* $\mathscr{L}$, and to H as the *order of* $\mathscr{L}$. Intuitively, $\underset{\sim}{0}$ is to represent falsehood and $\underset{\sim}{1}$ truth. The elements of $\mathscr{F} \cap \mathscr{C}$ and $\mathscr{F} \cap \mathscr{Q}$ will be called the *connectives* and the *quantifiers* of $\mathscr{L}$, respectively. We let $\|\mathscr{L}\|$ denote the *cardinal of* $\mathscr{L}$ and it is defined as

$$\|\mathscr{L}\| = \omega \cup |\pi| \cup |\kappa| \cup |\mathscr{S}_0| \cup |\mathscr{F}| \quad .$$

The following symbols will be associated with the continuous logic $\mathscr{L}$:

a denumerable list of *variables* v_n, $n < \omega$;

a π-termed sequence of *predicate symbols* P_ξ, $\xi < \pi$;

a κ-termed sequence of *constant symbols* c_ζ, $\zeta < \kappa$;

the *identity symbol* $\equiv$;

parentheses and comma .

We shall assume hereafter that

$$\mathcal{L} = (\langle \tau, \kappa \rangle, X, \mathcal{F}, \underset{\sim}{0}, \underset{\sim}{1}, H)$$

is an arbitrary but fixed continuous logic, and that $\pi = \mathcal{D}\tau$. We shall occasionally wish to vary some but not all of the terms of the sextuple $\mathcal{L}$. The new continuous logics so obtained will be denoted by the symbol $\mathcal{L}$ followed by those terms of $\mathcal{L}$ which are to be changed indicated between a pair of parentheses. For example, we shall write

$$\mathcal{L}(\mathcal{G}) = (\langle \tau, \kappa \rangle, X, \mathcal{G}, \underset{\sim}{0}, \underset{\sim}{1}, H) \quad .$$

The *dual logic of* $\mathcal{L}$ shall be written

$$\mathcal{L}(\underset{\sim}{1}, \underset{\sim}{0}, \breve{H}) = (\langle \tau, \kappa \rangle, X, \mathcal{F}, \underset{\sim}{1}, \underset{\sim}{0}, \breve{H}) \quad .$$

It is clear that the dual of a continuous logic is always a continuous logic. Notice that if H' is the identity relation on X, then $\mathcal{L}(H')$ is always a continuous logic. As a special case, we shall often vary the similarity type by increasing κ to $\kappa + \mu$, and in this case we shall adopt the briefer notation

$$\mathcal{L}(\mu) = \mathcal{L}(\langle \tau, \kappa + \mu \rangle) = (\langle \tau, \kappa + \mu \rangle, X, \mathcal{F}, \underset{\sim}{0}, \underset{\sim}{1}, H) \quad .$$

Although $\mathcal{L}$ is held fixed for the purpose of simplifying notation, we shall frequently use the fact that the results we obtain for $\mathcal{L}$ also hold for any other continuous logic $\mathcal{L}'$, and in particular for $\mathcal{L}(\mu)$.

Let us pause to discuss the part played by the order relation, H, in this monograph. It is perhaps more natural to first consider, instead of the above notion of a continuous logic, the simpler notion of an "unordered continuous logic,"

$$(\langle \tau, \kappa \rangle, X, \mathcal{F}, \underset{\sim}{0}, \underset{\sim}{1})$$

where $\langle \tau, \kappa \rangle$ is a similarity type, X is a compact Hausdorff space, $\mathcal{F} \subset \mathcal{C} \cup \mathcal{Q}$, $\mathcal{F}$ is closed under composition, and $\underset{\sim}{0}, \underset{\sim}{1}$ are two distinct points of X. In a large part of this monograph, including all of Chapters 4, 5, and 6, there is no mention at all of the relation H, and we could simply have used unordered continuous logics instead of continuous logics. On the other hand, the relation H is very important in Chapter 7. To make our exposition uniform, we have chosen to restrict our attention throughout to continuous logics $\mathcal{L}$, even when an unordered continuous logic would be sufficient. Because of the following lemma, the mathematical content is the same whether we use continuous logics or unordered continuous logics when the relation H is not involved in the discussion.

LEMMA 2.1.1. A quintuple

$$(\langle \tau, \kappa \rangle, X, \mathcal{F}, \underset{\sim}{0}, \underset{\sim}{1})$$

is an unordered continuous logic if and only if there exists an H such that the sextuple

$$(\langle \tau, \kappa \rangle, X, \mathcal{F}, \underset{\sim}{0}, \underset{\sim}{1}, H)$$

is a continuous logic.

PROOF. This follows from the remark in Section 1.4 that $(X, \underset{\sim}{0}, \underset{\sim}{1})$ is ordered by the identity relation, H, over X.

2.2. Formulas

The formulas of $\mathcal{L}$ will be defined to be certain finite sequences of symbols, connectives, and quantifiers. Arbitrary formulas will be constructed out of simpler formulas in a recursive fashion.

By an *equation (of $\mathcal{L}$)* we mean a 3-termed sequence

$$x_1 \equiv x_2 \quad ,$$

where x_1 and x_2 are either variables or constant symbols. By an *atomic formula (of $\mathcal{L}$)* we mean either an equation or a sequence

$$P_\xi(x_1, \ldots, x_{\tau(\xi)}) \quad ,$$

where $\xi < \pi$ and $x_1, \ldots, x_{\tau(\xi)}$ are either variables or constant symbols. We let Λ be the set of all atomic formulas (of $\mathcal{L}$).

The set Φ of all *formulas* (of $\mathcal{L}$) is defined as the least set Γ such that

2.2.1. $\Lambda \subset \Gamma$;

2.2.2. if $f \in \mathcal{C}_n \cap \mathcal{F}$ and $\varphi_1, \ldots, \varphi_n \in \Gamma$, then $f(\varphi_1, \ldots, \varphi_n) \in \Gamma$;

2.2.3. if $q \in \mathfrak{Q} \cap \mathcal{F}$, $\varphi \in \Gamma$, and $m < \omega$, then $(qv_m)\varphi \in \Gamma$.

When more than one continuous logic is involved in the same discussion, we shall sometimes write $\Phi_{\mathcal{L}}$ for the set Φ of formulas of $\mathcal{L}$, and $\Lambda_{\mathcal{L}}$ for the set Λ of atomic formulas of $\mathcal{L}$. If $\mathcal{G} \subset \mathcal{F}$ and $\Delta \subset \Phi$, we shall denote by $\mathcal{G}(\Delta)$, or by $\mathcal{G}_{\mathcal{L}}(\Delta)$ if we wish to indicate $\mathcal{L}$, the least set Γ of formulas of $\mathcal{L}$ such that each of the conditions 2.2.1, 2.2.2, 2.2.3 hold with $\mathcal{G}$, Δ in place of $\mathcal{F}$, Λ. Thus, for instance, we have $\Phi = \mathcal{F}(\Lambda)$. Notice that for all $\mathcal{G}$ and Δ,

$$\Delta \subset \mathcal{G}(\Delta) = \mathcal{G}(\mathcal{G}(\Delta)) \subset \Phi \quad .$$

LEMMA 2.2.4. Whenever $\Delta \subset \Phi$ and $\mathcal{G} \subset \mathcal{F}$, we have

.1) $|\mathcal{G}(\Delta)| \leq \omega \cup |\mathcal{G}| \cup |\Delta|$.

Moreover,

.2) $|\Phi| = \omega \cup |\mathcal{F}| \cup |\pi| \cup |\kappa|$.

PROOF. Since any formula $\varphi \in \mathcal{G}(\Delta)$ is a finite concatenation of members of $\mathcal{G}$, Δ, and variables, parentheses, and commas, the condition .1) holds. It is easily seen that

$$|\Lambda| = \omega \cup |\pi| \cup |\kappa| \quad ,$$

and that $|\Phi|$ is at least as large as $|\mathcal{F}|$ and $|\Lambda|$. therefore $|\Phi| = |\mathcal{F}| \cup |\Lambda|$ and .2) follows.

In connection with 2.2.4, we remark that $|\mathcal{G}(\Delta)|$ is finite if and only if either $\Delta = 0$, or else $\Delta \in S_\omega(\Phi)$ and $\mathcal{G} \in S_\omega(\mathcal{C}_0)$. When $|\mathcal{G}(\Delta)|$ is infinite, the inequality in .1) may be replaced by an identity. It follows from .2) that

$$\|\mathcal{L}\| = |\Phi| \cup |\mathcal{S}_0| \quad .$$

We shall say that a symbol, or a member of $\mathcal{F}$, occurs in a formula φ if it belongs to the range of the finite sequence φ. Thus only finitely many symbols and members of $\mathcal{F}$ occur in a given formula φ.

The set $v(\varphi)$ of free variables of a formula $\varphi \in \Phi$ may be defined recursively as follows:

2.2.5. if $\varphi \in \Lambda$, then $v(\varphi)$ is the set of all variables which occur in φ ;

2.2.6. if $f \in \mathcal{F} \cap \mathcal{C}_n$ and $\varphi = f(\psi_1, \ldots, \psi_n)$, then $v(\varphi) = v(\psi_1) \cup \ldots \cup v(\psi_n)$;

2.2.7. if $q \in \mathcal{Q} \cap \mathcal{F}$ and $\varphi = (qv_m)\psi$, then $v(\varphi) = v(\psi) - \{v_m\}$.

A formula φ is said to be a *sentence* if $v(\varphi) = 0$, and we shall denote the set of all sentences of $\mathcal{L}$ by Σ, or if necessary by $\Sigma_{\mathcal{L}}$.

EXERCISE 2A. Let $<\tau', \kappa'>$ be a type with $\tau \subset \tau'$ and $\kappa \leq \kappa'$, let $\mathcal{F} \subset \mathcal{F}' \subset \mathcal{C} \cup \mathcal{Q}$, and let $\mathcal{L}' = (<\tau', \kappa'>, X, \mathcal{F}', \underset{\sim}{0}, \underset{\sim}{1}, H)$. Then $\Lambda_{\mathcal{L}} \subset \Lambda_{\mathcal{L}'}$, $\Phi_{\mathcal{L}} \subset \Phi_{\mathcal{L}'}$, and $\Sigma_{\mathcal{L}} \subset \Sigma_{\mathcal{L}'}$. In particular, for any ordinal μ we have $\Lambda_{\mathcal{L}} \subset \Lambda_{\mathcal{L}(\mu)}$, $\Phi_{\mathcal{L}} \subset \Phi_{\mathcal{L}(\mu)}$, and $\Sigma_{\mathcal{L}} \subset \Sigma_{\mathcal{L}(\mu)}$; if η is a non-zero limit ordinal, then $\Lambda_{\mathcal{L}(\eta)} = \bigcup_{\mu < \eta} \Lambda_{\mathcal{L}(\mu)}$, $\Phi_{\mathcal{L}(\eta)} = \bigcup_{\mu < \eta} \Phi_{\mathcal{L}(\mu)}$, and $\Sigma_{\mathcal{L}(\eta)} = \bigcup_{\mu < \eta} \Sigma_{\mathcal{L}(\mu)}$.

2.3. Two-valued logic

In case X is the two point-space $\{0, 1\}$ with the discrete topology, the continuous logics with value space X are the ordinary two-valued logics. Among the familiar connectives and quantifiers of two-valued logic we have the following:

the *identity function* $\mathrm{id}(x) = x$;

the *negation* $\neg$ defined by $\neg(x) = 1 - x$;

the *conjunction* $\&$ defined by $\&(x, y) = \min(x, y)$;

the *disjunction* $\vee$ defined by $\vee(x, y) = \max(x, y)$;

the *existential quantifier* $\exists$ defined by $\exists(Y) = \sup(Y)$;

the *universal quantifier* $\forall$ defined by $\forall(Y) = \inf(Y)$.

There are a number of choices for the connectives and quantifiers, all of which lead to what may be called classical two-valued logics, and these logics are equivalent to each other in the sense that there is a natural way to translate the formulas of one into formulas of the other in such a way that the intended interpretation is preserved. To fix ideas, we shall agree to understand by the *classical two-valued logic of* type $< \tau, \kappa >$ the logic

$$\ell = (< \tau, \kappa >, \{\underset{\sim}{0}, \underset{\sim}{1}\}, \mathscr{F}_\ell, \underset{\sim}{0}, \underset{\sim}{1}, H_\ell) \quad ,$$

where $X = \{0, 1\}$, $\underset{\sim}{0} = 0$, $\underset{\sim}{1} = 1$, $\mathscr{F}_\ell \cap \mathscr{C}$ is the closure of $\{id, \&, \curlyvee, \neg\}$ under composition, $\mathscr{F}_\ell \cap \mathscr{Q} = \{\exists, \forall\}$, and H_ℓ is the natural ordering on $\{0, 1\}$.

We see that $\&$, $\curlyvee$, and id are H_ℓ-preserving connectives, and $\forall$ is an H_ℓ-preserving quantifier. On the other hand, $\neg$ and $\exists$ are not H_ℓ-preserving. Similarly, $\&$, $\curlyvee$, id, and $\exists$ are $\breve{H}_\ell$-preserving, but $\neg$ and $\forall$ are not $\breve{H}_\ell$-preserving. Both $\exists$ and $\forall$ are H_ℓ^*-preserving and $\breve{H}_\ell^*$-preserving. The dual of the logic ℓ, namely, the logic

$$\ell(\underset{\sim}{1}, \underset{\sim}{0}, \breve{H}_\ell) = (< \tau, \kappa >, \{0, 1\}, \mathscr{F}_\ell, \underset{\sim}{1}, \underset{\sim}{0}, \breve{H}_\ell)$$

is in many respects anti-isomorphic to ℓ. That is to say, many true statements about ℓ become true statements about $\ell(\underset{\sim}{1}, \underset{\sim}{0}, \breve{H}_\ell)$ when the roles of $\&$ and $\curlyvee$, $\exists$ and $\forall$, $\underset{\sim}{0}$ and $\underset{\sim}{1}$, and H_ℓ and $\breve{H}_\ell$ are interchanged.

2.4. Sets of connectives and quantifiers

In this section we shall consider special properties of connectives and quantifiers. Sets satisfying these properties in the model theory for a general continuous logic $\mathscr{L}$ will correspond to the sets $\{id, \neg\}$, $\{id\}$, $\{\&\}$, and $\{\exists\}$ in the logic ℓ.

2.4.1. $\mathscr{T}$ is said to be a t-set, or a *transfer set* (of $\mathscr{L}$), if $\mathscr{T} \subset \mathscr{F} \cap \mathscr{C}_1$ and, for all $x, y \in X$ with $x \neq y$, there is a $t \in \mathscr{T}$ such that $t(x) = \underset{\sim}{1}$ and $t(y) \neq \underset{\sim}{1}$.

It is easily seen that the set $\{id, \neg\}$ is a t-set of ℓ and of $\ell(\underset{\sim}{1}, \underset{\sim}{0}, \breve{H}_\ell)$.

2.4.2. $\mathscr{H}$ is said to be an H-set, or *transfer set for* H (of $\mathscr{L}$), if $\mathscr{H} \subset \mathscr{F} \cap \mathscr{C}_1 \cap \mathscr{C}_H$ and, whenever $< x, y > \in X^2 - H$, there is a $t \in \mathscr{H}$ such that $t(x) = \underset{\sim}{1}$ and $t(y) \neq \underset{\sim}{1}$.

Note that $\{id\}$ is an H_ℓ-set of ℓ and an $\breve{H}_\ell$-set of $\ell(\underset{\sim}{1}, \underset{\sim}{0}, \breve{H}_\ell)$. In general, if H is the identity relation on X, then $\mathscr{T}$ is a t-set if and only if it is an H-set.

2.4.3. $\mathscr{K}$ is said to be a k-set, or *conjunction set* (of $\mathscr{L}$), if $\mathscr{K} \subset \mathscr{F} \cap \mathscr{C}_2$, if $k(\underset{\sim}{1}, \underset{\sim}{1}) = \underset{\sim}{1}$ for all $k \in \mathscr{K}$, and if whenever $\underset{\sim}{1} \in V \in \mathscr{S}_0$ there exists $k \in \mathscr{K}$ such that $k^{\cup}[\underset{\sim}{1}] \subset V^2$.

Thus $\{\&\}$ is a k-set of ℓ, and $\{\vee\}$ is a k-set of $\ell(\underset{\sim}{1}, \underset{\sim}{0}, \breve{H}_\ell)$.

2.4.4. $\mathscr{E}$ is said to be an e-set, or *existential set* (of $\mathscr{L}$), if $\mathscr{E} \subset \mathscr{F} \cap \mathscr{Q}$ and, for all $Y \in X^*$, we have

$$\{e(Y) : e \in \mathscr{E}\} = \{\underset{\sim}{1}\} \quad \text{if and only if} \quad \underset{\sim}{1} \in \bar{Y} \quad .$$

For example, $\{\exists\}$ is an e-set of ℓ and $\{\forall\}$ is an e-set of $\ell(\underset{\sim}{1}, \underset{\sim}{0}, \breve{H}_\ell)$. Notice that any t-set, k-set, or e-set of $\mathscr{L}$ is non-empty. Since $H \neq X^2$ (see Ex. 1L), any H-set of $\mathscr{L}$ is also non-empty. If $\mathscr{F} \subset \mathscr{G} \subset \mathscr{C} \cup \mathscr{Q}$, then any t-set, H-set, k-set, or e-set of $\mathscr{L}$ is obviously a t-set, H-set, k-set, or e-set, respectively of $\mathscr{L}(\mathscr{G})$.

THEOREM 2.4.5. .1) Let $\mathscr{T}$ be a t-set. Then whenever $x \in V \in \mathscr{S}$, there is a $\mathscr{T}_0 \in S_\omega(\mathscr{T})$ such that

$$\{t(x): \ t \in \mathscr{T}_0\} = \{\underset{\sim}{1}\} \quad ,$$

and for all $y \in X$,

$$\{t(y): \ t \in \mathscr{T}_0\} = \{\underset{\sim}{1}\} \quad \text{implies} \quad y \in V \quad .$$

.2) Let $\mathscr{H}$ be an H-set. Then whenever $x \in V \in \mathscr{S}$, there is an $\mathscr{H}_0 \in S_\omega(\mathscr{H})$ such that

$$\{t(x): \ t \in \mathscr{H}_0\} = \{\underset{\sim}{1}\} \quad ,$$

and for all $y \in X$,

$$\{t(y): \ t \in \mathscr{H}_0\} = \{\underset{\sim}{1}\} \quad \text{implies} \quad y \in H[V] \quad .$$

.3) Let $\mathscr{E}$ be an e-set. Then whenever $\underset{\sim}{1} \in V \in \mathscr{S}$, there is an $\mathscr{E}_0 \in S_\omega(\mathscr{E})$ such that for all $Y \in X^*$,

$$\{e(Y): \ e \in \mathscr{E}_0\} = \{\underset{\sim}{1}\} \quad \text{implies} \quad Y \cap V \neq 0 \quad .$$

PROOF. .1) Let $x \in V \in \mathscr{S}$ and let $\mathscr{T}_1 = \{t \in \mathscr{T}: t(x) = \underset{\sim}{1}\}$. Then the set

$$\{t^{\cup}[X - \{\underset{\sim}{1}\}]: \ t \in \mathscr{T}_1\}$$

is an open covering of the compact set $X - V$. This open covering has a finite subcovering

$$\{t^{\smile}[X - \{\underset{\sim}{1}\}]: \quad t \in \mathcal{T}_0\} \quad ,$$

where $\mathcal{T}_0 \in S_\omega(\mathcal{T}_1)$. Then $t(x) = \underset{\sim}{1}$ for all $t \in \mathcal{T}_0$, and whenever $y \in X - V$, there is a $t \in \mathcal{T}_0$ such that $t(y) \neq \underset{\sim}{1}$. Hence $\mathcal{T}_0$ has the desired properties.

The proof of .2) is an obvious modification of the proof of .1). Note that $H[V] \in \mathcal{S}$ whenever $V \in \mathcal{S}$.

.3) Let $\underset{\sim}{1} \in V \in \mathcal{S}$. The set

$$\{e^{\smile}[X - \{\underset{\sim}{1}\}]: \quad e \in \mathcal{E}\}$$

is an open covering of the closed set $(X - V)^*$ in the space X^*. Since X^* is compact (Theorem 1.5.8), this open covering has a finite subcovering

$$\{e^{\smile}[X - \{\underset{\sim}{1}\}]: \quad e \in \mathcal{E}_0\}$$

where $\mathcal{E}_0 \in S_\omega(\mathcal{E})$. Hence if $Y \in X^*$ and $e(Y) = \underset{\sim}{1}$ for all $e \in \mathcal{E}_0$, then $Y \cap V \neq 0$. This completes the proof.

In the foregoing theorem, conditions .1), .2), and .3) are "local" conditions in which we have a finite set of functions corresponding to each open neighborhood. On the other hand, the relevant definitions 2.4.1, 2.4.2, and 2.4.4 of t-set, H-set, and e-set may be thought of as "pointwise" conditions. The definitions could have been formulated locally in an equivalent way, but such formulations would not be as simple. The definition 2.4.3 of k-set is a "strong local" condition in which a single function corresponds to each neighborhood; the obvious "pointwise" and "local" conditions prove to be too weak for our intended use of the notion of a k-set.

We conclude this section with two lemmas.

LEMMA 2.4.6. Let $\mathcal{E}$ be an e-set and let $e \in \mathcal{E}$. Then whenever $\underset{\sim}{1} \in V \in \mathcal{S}$, there exists $\underset{\sim}{1} \in U \in \mathcal{S}$ such that for all $Y \in X^*$, $Y \cap U \neq 0$ implies $e(Y) \in V$.

PROOF. Let $\underset{\sim}{1} \in V \in \mathcal{S}$. Suppose to the contrary that whenever $\underset{\sim}{1} \in U \in \mathcal{S}$, there exists $Y \in X^*$ such that $Y \cap U \neq 0$ and $e(Y) \notin V$. Let $\mathcal{S}' = \{U: \underset{\sim}{1} \in U \in \mathcal{S}\}$ and let $I = S_\omega(\mathcal{S}')$. For each $i \in I$, define

$$F(i) = \{Y\colon Y \cap U \neq 0 \quad \text{for each} \quad U \in i \quad \text{and} \quad e(Y) \notin V\} \quad .$$

Clearly $F\colon \; I \to (X^*)^*$ and for all $i_1, i_2 \in I$,

$$F(i_1 \cup i_2) \subset F(i_1) \cap F(i_2) \quad .$$

Let $f \in \Pi_{i \in I} F(i)$ and let D be a regular ultrafilter over I. By 1.5.7*,

$$F(\{U\}) \cap W' \neq 0$$

for all $U \in \mathscr{S}'$ and $W' \in \mathscr{S}^*$ such that $D^*\text{-lim}\, f \in W'$. Since e is continuous and each $e(f(i)) \notin V$, by 1.5.6 and 1.5.2, $e(D^*\text{-lim}\, f) \notin V$. Since $e \in \mathscr{E}$ and $\mathscr{E}$ is an e-set, this means $\underset{\sim}{1} \notin \overline{D^*\text{-lim}\, f}$. By the normality of X, let $U, W \in \mathscr{S}$ be such that $\underset{\sim}{1} \in U$, $\overline{D^*\text{-lim}\, f} \subset W$, and $U \cap W = 0$. Let $W' \in \mathscr{S}^*$ be defined as $W' = \{Y \in X^*\colon \bar{Y} \subset W\}$. Note that $D^*\text{-lim}\, f \in W'$. Hence, there exists $Y \in X^*$ such that $Y \cap U \neq 0$ and $\bar{Y} \subset W$. This is clearly impossible. The lemma is proved.

LEMMA 2.4.7. Let $\mathscr{K}$ be a k-set and let $\underset{\sim}{1} \in V \in \mathscr{S}$. Then for every integer $n \geq 2$ there exists a function $k_n \in \mathscr{C}_n$ obtained by composition from members of $\mathscr{K}$ such that

$$k_n(\underset{\sim}{1}, \underset{\sim}{1}, \ldots, \underset{\sim}{1}) = \underset{\sim}{1} \quad \text{and} \quad \overset{\cup}{k_n}[\underset{\sim}{1}] \subset V^n \quad .$$

PROOF. Let $k \in \mathscr{K}$ be such that $\overset{\cup}{k}[\underset{\sim}{1}] \subset V^2$. We define k_n, for $n \geq 2$, inductively as follows:

$$k_2 = k \quad ;$$

$$k_{n+1}(x_1, \ldots, x_{n+1}) = k'(k_n(x_1, \ldots, x_n), x_{n+1}) \quad \text{where}$$

$$k' \in \mathscr{K} \text{ and } \overset{\cup}{k'}[\underset{\sim}{1}] \subset (V - k_n[X^n - V^n])^2.$$

We can show that for each $n \geq 2$,

$$k_n \in \mathscr{C}_n, \quad k_n(\underset{\sim}{1}, \ldots, \underset{\sim}{1}) = \underset{\sim}{1}, \quad \text{and} \quad \overset{\cup}{k_n}[\underset{\sim}{1}] \subset V^n \quad .$$

The lemma is proved.

We remark that in case $\mathscr{K} \subset \mathscr{C}_H$, then $k_n \in \mathscr{C}_H$.

EXERCISE 2B. Give a detailed proof of Lemma 2.4.7.

2.5. Examples

EXAMPLE 2.5.1. The classical two-valued logic ℓ has the t-set $\{id, \neg\}$, the H_ℓ-set $\{id\}$, the k-set $\{\&\}$ which contain only

H_ℓ-preserving connectives, and the e-set $\{\exists\}$ which contains only $\check{H}_\ell$-preserving quantifiers.

EXAMPLE 2.5.2. Let X be the closed real unit interval [0, 1] with the usual topology. Take $\underset{\sim}{0} = 0$ and $\underset{\sim}{1} = 1$. Let $\mathscr{F}$ be defined as follows:

$\mathscr{F} \cap \mathscr{C}$ is the closure of $\{(1-x), \min(1, x+y)\}$ under composition;

$\mathscr{F} \cap \mathscr{Q} = \{\sup\}$.

Let H be the natural ordering on [0, 1]. This continuous logic $\mathscr{L} = (\langle \tau, \kappa \rangle, X, \mathscr{F}, \underset{\sim}{0}, \underset{\sim}{1}, H)$ is the classical infinite real valued logic. In what follows we shall show that $\mathscr{L}$ has a t-set, an H-set, a k-set, and an e-set.

Clearly $\{\sup\}$ is an e-set of $\mathscr{L}$. Furthermore, $\sup \in \mathscr{Q}_{\check{H}}$.

The function $\min(x, y)$ can be obtained from $(1-x)$ and $\min(1, x+y)$ as follows:

$$\min(x, y) = 1 - \min(1, (1-y) + (1-\min(1, x + (1-y)))) \quad .$$

Thus $\min(x, y) \in \mathscr{F} \cap \mathscr{C}$ and $\{\min(x, y)\}$ is k-set of $\mathscr{L}$. Notice that $\min(x, y) \in \mathscr{C}_H$.

To obtain the t-set and H-set of $\mathscr{L}$, we first show that for each rational $x \in X$, there are two functions $f_x, g_x \in \mathscr{F} \cap \mathscr{C}_1 \cap \mathscr{C}_H$ such that $\check{f}_x[1] = [x, 1]$ and $\check{g}_x[0] = [0, x]$.

To simplify the proceedings, we introduce the following temporary notation:

$$\bar{x} = 1 - x \; ;$$
$$x \dot{+} y = \min(1, x+y) \; ;$$
$$x \cdot y = (\bar{x} \dot{+} \bar{y})^- \; ;$$
$$n(x) = x \dot{+} x \dot{+} \dots \dot{+} x \qquad \text{n-times} \; ;$$
$$(x)^n = x \cdot x \cdot \dots \cdot x \qquad \text{n-times} \; .$$

We see that both $x \dot{+} y$, $x \cdot y \in \mathscr{F} \cap \mathscr{C}_H$. We first define $f_0(x) = x \dot{+} \bar{x}$, $f_1(x) = (\bar{x})^-$. Let x be a rational of the form p/q where $1 \leq p \leq (q-1)$ and $2 \leq q$. Suppose that $q = 2$, then we define

$$f_{\frac{1}{2}}(y) = 2(y) \quad .$$

Suppose that $f_{p'/q'}$ has been found for all $q' < q$ and all p' such that $1 \leq p' \leq (q'-1)$. We find $f_{p/q}$ by induction on p. If $p = 1$, then

$$f_{1/q}(y) = q(y) \quad .$$

Suppose that $f_{p'/q}$ has been found for all p' such that $1 \leq p' < p$. If p and q are not relatively prime, then $f_{p/q}$ has already been found. If p and q are relatively prime, then there exist numbers m and n such that

$$[m(p/q)]^n = p'/q \text{ for some } p' \text{ such that } 1 \leq p' < p \quad .$$

Hence, define

$$f_{p/q}(y) = f_{p'/q}([m(y)]^n) \quad .$$

EXERCISE 2C. Find the functions $g_x \in \mathscr{F} \cap \mathscr{C}_H$.

Continuing with our example, the set $\{f_x\colon\ x \text{ rational}\}$ is an H-set of $\mathscr{L}$, and the set

$$\{f_x\colon\ x \text{ rational}\} \cup \{1 - g_x\colon\ x \text{ rational}\}$$

is a t-set of $\mathscr{L}$.

EXAMPLE 2.5.3. Let X be the closed parallelogram in the complex plane with vertices at 0, $1 + 2i$, $2 + i$, and $3 + 3i$. Let $\mathscr{S}$ be the usual topology on X, and let $\underset{\sim}{0} = 0$ and $\underset{\sim}{1} = 3 + 3i$. Let

$$H = \{< a + bi, c + di > \in X^2\colon\ a \leq c \text{ and } b \leq d\} \quad .$$

For each $x \in X$, let t_x be the unique function in $\mathscr{C}_1$ such that $\breve{t}_x[\underset{\sim}{1}] = H[x]$, and such that the restriction of t_x to any line through $\underset{\sim}{1}$ is a rigid translation along that line. Similarly, let u_x be the unique function in $\mathscr{C}_1$ such that $\breve{u}_x[\underset{\sim}{0}] = \breve{H}[x]$, and such that the restriction of u_x along any line through $\underset{\sim}{0}$ is a rigid translation along that line. Define $t_x'(y) = \underset{\sim}{1} - u_x(y)$. Let $\mathscr{F} \cap \mathscr{C}$ be the closure under composition of the following functions:

$\min(x, y)$ taken with respect to H ;

t_x where $x = a + bi \in X$ and a, b are rational ;

t_x' where $x = a + bi \in X$ and a, b are rational .

Let $\mathscr{F} \cap \mathscr{Q} = \{\sup(Y)\}$, where sup is taken with respect to H. It is easy to check that

$$\mathcal{L} = (\langle \tau, \kappa \rangle, X, \mathcal{F}, \underset{\sim}{0}, \underset{\sim}{1}, H)$$

is a continuous logic, that the set

$$\mathcal{H} = \{t_x: \ x = a + bi \in X \ \text{and} \ a, b \ \text{rational}\}$$

is an H-set of $\mathcal{L}$, that

$$\mathcal{H} \cup \{t_x^!: \ x = a + bi \in X \ \text{and} \ a, b \ \text{rational}\}$$

is a t-set of $\mathcal{L}$, that the set {min} is a k-set of $\mathcal{L}$ with min $\in \mathcal{C}_H$, and that the set {sup} is an e-set of $\mathcal{L}$ with sup $\in \mathcal{Q}_H$.

The three preceding examples have the particularly pleasant properties, which we shall not insist upon in general, that $\mathcal{F}$ is countable and that the k-set and e-set each has exactly one member.

The following exercises show some limitations on the notion of continuous logic.

EXERCISE 2D*. Consider the classical two-valued propositional logic $\mathcal{P}$ with propositional variables (zero-placed predicates) $p_0, p_1, \ldots,$ and connectives $\neg, \&, \vee$. If φ is a formula of $\mathcal{P}$, let (φ) be the set of all formulas which are logically equivalent to φ. Let

$$X = \{(\varphi) : \varphi \text{ is a formula of } \mathcal{P}\} .$$

The *Lindenbaum algebra* on X is the algebra $\langle X, \neg, \&, \vee \rangle$ where $\neg, \&, \vee$ are the operations on X defined by

$$\neg(\varphi) = (\neg\varphi), \quad (\varphi)\&(\psi) = (\varphi\&\psi), \quad (\varphi)\vee(\psi) = (\varphi\vee\psi) .$$

It is the free Boolean algebra with the generators $(p_0), (p_1), \ldots$. Prove that there is no compact Hausdorff topology on the set X such that the functions $\neg, \&, \vee$ are continuous. Thus; the Lindenbaum algebra "cannot be made into a continuous logic."

EXERCISE 2E*. If we consider the intuitionistic logic, e.g., as developed in Kleene [1952, p. 82], we obtain the <u>Brouwerian algebra</u> $\langle X, \neg, \&, \vee, \supset \rangle$ on the set X of equivalence classes of formulas. Prove that the Brouwerian algebra cannot be made into a continuous logic in the sense of the preceding Exercise 2D.

Hint for Exercises 2D and 2E: For each formula φ, the sets

$\{(\psi): (\psi) \leq (\varphi)\}$ and $\{(\psi): (\varphi) \leq (\psi)\}$ must be closed in any Hausdorff topology for which the operations are continuous.

EXERCISE 2F*. The intuitionistic propositional logic can be interpreted as a many-valued logic whose set of truth values is the set X of all open subsets of the euclidean plane E^2. The functions &, ⅋, ¬ and ⊃ on X are defined by:

$$x \mathbin{\&} y = x \cap y \ ;$$

$$x \mathbin{⅋} y = x \cup y \ ;$$

$$(\neg y) = \text{interior of } (E^2 - y) \ ;$$

$$(x \supset y) = \text{interior of } ((E^2 - x) \cup y) \ .$$

Prove that there is no compact Hausdorff topology on X for which ¬ and ⊃ are continuous. Hence, there is no continuous logic $\mathscr{L}$ with X as value space and with the functions ¬ and ⊃ belonging to $\mathscr{F} \cap \mathscr{C}$.

Hint: Show that for each $a \in E^2$, the sets $\{x \in X: a \in x\}$ and $\{x \in X: a \in \neg x\}$ are open.

2.6. Some existence theorems

In the chapters to follow, we shall frequently assume, as a hypothesis to a theorem, that (1) $\mathscr{L}$ has a t-set, a k-set, and an e-set, or else (2) $\mathscr{L}$ has an H-set, a k-set of H-preserving connectives, and an e-set of $\breve{H}$-preserving quantifiers. Thus each of the three examples of 2.5 above satisfies (1) and (2). In this section, we shall give some sufficient conditions for a continuous logic $\mathscr{L}$ to have the property (1) or the property (2). While we shall not attempt to characterize all such logics $\mathscr{L}$, we shall at least establish that the results to be obtained in later chapters are far from vacuous. For convenience, throughout this section we shall only consider continuous logics

$$\mathscr{L} = (\langle \tau, \kappa \rangle, X, \mathscr{F}, \underset{\sim}{0}, \underset{\sim}{1}, H) \text{ with } \mathscr{F} = \mathscr{C} \cup \mathscr{Q} \ .$$

THEOREM 2.6.1. .1) Suppose X is a compact Boolean space, i.e., a compact Hausdorff space for which the closed open sets form a base. Then $\mathscr{L}$ has a t-set, a k-set, and an e-set.

.2) Suppose that $\underset{\sim}{0}\ H\ \underset{\sim}{1}$, and whenever $< x, y > \in X^2 - H$, there is a closed open set W such that $x \in W$, $y \notin W$, and $H[W] = W$. Then $\mathcal{L}$ has an H-set $\mathcal{H}$, a k-set $\mathcal{K} \subset \mathcal{C}_H$, and an e-set $\mathcal{E} \subset \mathcal{Q}_H^{\vee}$.

PROOF. .1) We first prove that $\mathcal{L}$ has a t-set. Suppose x, $y \in X$ and $x \neq y$. Then there is a closed open set W such that $x \in W$ and $y \notin W$. Let

$$t_{x\,y} = (W \times \{\underset{\sim}{1}\}) \cup ((X-W) \times \{\underset{\sim}{0}\}) \quad .$$

Then $t_{x\,y}$ is continuous, maps x into $\underset{\sim}{1}$ and y into $\underset{\sim}{0}$. Hence the set

$$\{t_{xy}\colon\ x, y \in X \text{ and } x \neq y\} \subset \mathcal{C}_1$$

is a t-set of $\mathcal{L}$.

To construct a k-set, let $\underset{\sim}{1} \in V \in \mathcal{S}_0$, The compact set $X - V$ has a covering consisting of closed open sets which do not contain $\underset{\sim}{1}$, and this covering has a finite subcovering $\{U_1, \ldots, U_n\}$. The set

$$W = X - (U_1 \cup \ldots \cup U_n)$$

is then a closed open set such that $\underset{\sim}{1} \in W \subset V$. Moreover, W^2 is a closed open set of X^2. The function

$$k_V = (W^2 \times \{\underset{\sim}{1}\}) \cup ((X^2 - W^2) \times \{\underset{\sim}{0}\})$$

is continuous and satisfies the conditions $k_V(\underset{\sim}{1}, \underset{\sim}{1}) = \underset{\sim}{1}$ and $\breve{k}_V[\underset{\sim}{1}] \subset V^2$. Hence the set

$$\{k_V\colon\ \underset{\sim}{1} \in V \in \mathcal{S}_0\} \subset \mathcal{C}_2$$

is a k-set of $\mathcal{L}$.

Suppose that $Y \in X^*$ and $\underset{\sim}{1} \notin \bar{Y}$. As in the preceding paragraph there is a closed open set W such that $\underset{\sim}{1} \in W \subset X - \bar{Y}$. Since

$$(X - W)^* = \{Z \in X^*\colon\ \bar{Z} \subset X - W\} \quad ,$$

the set $(X - W)^*$ is open in X^*; moreover, since

$$X^* - (X - W)^* = \{Z \in X^*\colon\ Z \cap W \neq 0\} \quad ,$$

$(X - W)^*$ is also closed in X^*. It follows that the function

$$e_Y = ((X^* - (X-W)^*) \times \{\underset{\sim}{1}\}) \cup ((X - W)^* \times \{\underset{\sim}{0}\})$$

belongs to $\mathcal{Q}$, and has the properties that $e_Y(Y) = 0$ and $e_Y(Z) = \underset{\sim}{1}$ whenever $\underset{\sim}{1} \in \bar{Z}$. Thus the set

$$\{e_Y: Y \in X^* \text{ and } 1 \notin \bar{Y}\} \subset \mathcal{Q}$$

is an e-set of $\mathcal{L}$.

The proof of .2) is similar except that the closed open sets W must be picked so that $H[W] = W$. We notice that this can always be done because $H[\underset{\sim}{1}] = \{\underset{\sim}{1}\}$ and

$$H[W_1 \cap \ldots \cap W_n] = W_1 \cap \ldots \cap W_n$$

whenever

$$H[W_1] = W_1, \ldots, H[W_n] = W_n .$$

The compact Boolean spaces referred to in 2.6.1, .1) are exactly the Stone spaces of Boolean algebras (or of Boolean rings with units; see Kelley [1955, pp. 168-169]). In particular, any finite Hausdorff space is a compact Boolean space. As examples of compact Hausdorff spaces X and relations H satisfying the hypotheses of 2.6.1, .2) we shall mention the following. Any finite subset X of the real closed interval $[0, 1]$ with $\underset{\sim}{0} = 0$, $\underset{\sim}{1} = 1$, and where H is the natural order. The Cantor discontinuum (see Kelley [1955, p. 165]) with the topology induced by the reals, $\underset{\sim}{0} = 0$, $\underset{\sim}{1} = 1$, and where H is the natural order.

The hypotheses of 2.6.1,.2) do not imply that X is a Boolean space; for example they are satisfied by every continuous logic $\mathcal{L}$ such that $\underset{\sim}{0}$ and $\underset{\sim}{1}$ are isolated points of X and

$$H = (\{\underset{\sim}{0}\} \times X) \cup (X \times \{\underset{\sim}{1}\}) \cup (X - \{\underset{\sim}{0}, \underset{\sim}{1}\})^2 .$$

Because of the special importance of the logics with finite value spaces, we state the following

COROLLARY 2.6.2. .1) If X is finite, then $\mathcal{L}$ has a t-set, a k-set, and an e-set.
.2) If X is finite and $\underset{\sim}{0}\, H\, \underset{\sim}{1}$, then $\mathcal{L}$ has an H-set, a k-set $\mathcal{K} \subset \mathcal{C}_H$, and en e-set $\mathcal{E} \subset \mathcal{Q}_H^{\omega}$.

PROOF. .1) follows from 2.6.1, .1) and the fact that any finite Hausdorff space is a compact Boolean space.

.2) follows from 2.6.1, .2) because under our hypotheses, if $< x, y > \in X^2 - H$, then the set $Y = H[x]$ is a closed open set of X such that $x \in Y$, $y \notin Y$, and $H[Y] = Y$.

By a *(non-trivial) arc* a *in* X we mean a continuous function $a\colon [0, 1] \to X$, where $[0, 1]$ is the closed real unit interval and such that $a(1) = \underset{\sim}{1}$ and $a(0) \neq \underset{\sim}{1}$.

THEOREM 2.6.3. If there is an arc a in X, then $\mathcal{L}$ has a t-set, a k-set, and an e-set.

PROOF. Let $x, y \in X$ and $x \neq y$. By Urysohn's Lemma there is a continuous function $f\colon X \to [0, 1]$ such that $f(x) = 1$ and $f(y) = 0$. If we define $t_{xy} = a \circ f$, then $t_{xy} \in C_1$, $t_{xy}(x) = \underset{\sim}{1}$, and $t_{xy}(y) \neq \underset{\sim}{1}$. Hence the set

$$\{t_{xy}\colon\ x, y \in X \text{ and } x \neq y\}$$

is a t-set of $\mathcal{L}$.

Let $\underset{\sim}{1} \in V \in \mathcal{S}_0$. By Urysohn's Lemma there is a continuous function $g\colon X^2 \to [0, 1]$ such that $g(\underset{\sim}{1}, \underset{\sim}{1}) = 1$ and $g[X^2 - V^2] = \{0\}$. Let $k_V = a \circ g$; then $k_V \in \mathcal{C}_2$, $k_V(\underset{\sim}{1}, \underset{\sim}{1}) = 1$, and $k^{\cup}[\underset{\sim}{1}] \subset V^2$. Hence the set

$$\{k_V\colon\ \underset{\sim}{1} \in V \in \mathcal{S}_0\}$$

is a k-set of $\mathcal{L}$.

As in the preceding paragraph, we again assume that $\underset{\sim}{1} \in V \in \mathcal{S}_0$. Then the sets

$$U = \{Z \in X^*\colon\ \underset{\sim}{1} \in \bar{Z}\}$$

and

$$W = \{Y \in X^*\colon\ Y \cap V = 0\}$$

are disjoint closed sets in the space X^*. Since the restriction of X^* to the closed sets of X^* is a compact Hausdorff space (see Theorem 1.5.8 and the discussion in Section 1.3), we may apply Urysohn's Lemma to obtain a continuous function h_0 on that Hausdorff space into the interval $[0, 1]$ such that $h_0[U] = \{1\}$ and $h_0[W] = \{0\}$. h_0 can be extended to a continuous function $h\colon X^* \to [0, 1]$ by setting $h(Y) = h_0(\bar{Y})$ for all $Y \in X^*$. Then $h[U] = \{1\}$ and $h[W] = \{0\}$. Let $e_V = a \circ h$, and let

$$\mathcal{E} = \{e_V\colon\ \underset{\sim}{1} \in V \in \mathcal{S}_0\} \quad .$$

Since $a(1) = \underset{\sim}{1}$ and $a(0) \neq \underset{\sim}{1}$, and whenever $\underset{\sim}{1} \notin \bar{Y}$, there is a $V \in \mathcal{S}_0$ such that $\underset{\sim}{1} \in V$ and $Y \cap V = 0$, it follows that $\mathcal{E}$ is an e-set of $\mathcal{L}$.

In both Theorems 2.6.1 and 2.6.3 it is not actually necessary that $\mathcal{F} = \mathcal{C} \cup \mathcal{Q}$. In fact all that is needed is that the connectives and quantifiers used in the proofs to make up the t-sets, k-sets, and e-sets belong to $\mathcal{F}$.

The proof of Theorem 2.6.3 suggests a considerable variety of examples of continuous logics $\mathcal{L}$ which have a t-set, a k-set, and an e-set. For instance, suppose the space X has a family a_i, $i \in I$, of non-trivial arcs. Then if for each two distinct points $x, y \in X$ there is a continuous function $f\colon X \to [0, 1]$ with $f(x) = 1$, $f(y) = 0$, and an $i \in I$ such that $a_i \circ f \in \mathcal{F}$, then $\mathcal{L}$ has a t-set. Similarly, k-sets and e-sets can be constructed which involve arcs a_i with the index i varying.

One could also state a part .2) for Theorem 2.6.3 in the spirit of part .2) of Theorem 2.6.1, and which involves the relation H. The hypotheses of such a statement would be complicated considerably by the fact that Urysohn's Lemma would no longer be applicable directly. We shall instead prove a much more general existence theorem which includes all the previous results of this section as special cases. Roughly speaking, the result, which follows below, states that if a continuous logic

$$\mathcal{L}' = (\langle \tau, \kappa \rangle, X', \mathcal{F}', \underset{\sim}{0}', \underset{\sim}{1}', H')$$

has a t-set, a k-set, and an e-set, then so does every continuous logic $\mathcal{L}$ which is in a certain topological relationship with $\mathcal{L}'$. We do not need to assume that $\mathcal{F}' = \mathcal{C}' \cup \mathcal{Q}'$, but we do assume that $\mathcal{F} = \mathcal{C} \cup \mathcal{Q}$.

THEOREM 2.6.4. Suppose that $\mathcal{L}$ and $\mathcal{L}'$ are continuous logics such that:

(a) $\mathcal{L}'$ has an H'-set $\mathcal{H}'$, a k-set $\mathcal{K}' \subset \mathcal{C}_{H'}$, and an e-set $\mathcal{E}' \subset \mathcal{Q}_{H'}$;

(b) whenever $\langle x, y \rangle \in X^2 - H$, there is a function $a_{xy} \in \mathcal{C}(X, X')$ such that $\langle a_{xy}(x), a_{xy}(y) \rangle \notin H'$, and whenever $u H v$, then $a_{xy}(u) H' a_{xy}(v)$;

(c) whenever $\underset{\sim}{1}' \in V' \in \mathcal{S}'$, there is a function $b_{V'} \in \mathcal{C}(X', X)$ such that $b_{V'}(\underset{\sim}{1}') = \underset{\sim}{1}$, $\breve{b}_{V'}[\underset{\sim}{1}] \subset V'$, and whenever $u'H'v'$, then $b_{V'}(u')Hb_{V'}(v')$.

Then $\mathcal{L}$ has an H-set $\mathcal{H}$, a k-set $\mathcal{K} \subset \mathcal{C}_H$, and an e-set $\mathcal{E} \subset \mathcal{Q}_{\breve{H}}$.

PROOF. Suppose that $< x, y > \in X^2 - H$. Then $< a_{xy}(x), a_{xy}(y) > \notin H'$, and hence there exists $t \in \mathcal{H}'$ such that $t(a_{xy}(x)) = \underset{\sim}{1}'$ and $t(a_{xy}(y)) \neq \underset{\sim}{1}'$. Let $V' = X' - \{t(a_{xy}(y))\}$ and define the function $t_{xy}: X \to X$ by

$$t_{xy} = b_{V'} \circ t \circ a_{xy} \quad .$$

Clearly $t_{xy} \in \mathcal{C}_1 \cap \mathcal{C}_H$ and $t_{xy}(x) = \underset{\sim}{1}$ and $t_{xy}(y) \neq \underset{\sim}{1}$. Hence the set

$$\mathcal{H} = \{t_{xy}: < x, y > \in X^2 - H\}$$

is an H-set of $\mathcal{L}$.

Before we continue with the proof, we first prove the dual to (c), namely

(c') whenever $\underset{\sim}{1} \in V \in \mathcal{S}$, there is a function $a_V \in \mathcal{C}(X, X')$ such that $a_V(\underset{\sim}{1}) = \underset{\sim}{1}'$, $\breve{a}_V[\underset{\sim}{1}'] \subset V$, and whenever $u H v$, then $a_V(u)H'a_V(v)$.

Let $\underset{\sim}{1} \in V \in \mathcal{S}$. Since $H[\underset{\sim}{1}] = \{\underset{\sim}{1}\}$, we have $< \underset{\sim}{1}, y > \in X^2 - H$ for all $y \in X - V$. As in the preceding paragraph, for each $y \in X - V$, pick $t_y \in \mathcal{H}'$ such that

$$(t_y \circ a_{\underset{\sim}{1}y})(\underset{\sim}{1}) = \underset{\sim}{1}' \quad \text{and} \quad (t_y \circ a_{\underset{\sim}{1}y})(y) \neq \underset{\sim}{1}' \quad .$$

Let $f_y = t_y \circ a_{\underset{\sim}{1}y}$, we see that $f_y \in \mathcal{C}(X, X')$ and $f_y(u)H'f_y(v)$ whenever $u H v$. Let

$$U_y = \breve{f}_y[X' - \{\underset{\sim}{1}'\}] \quad ,$$

and choose $W_y \in \mathcal{S}$ so that

$$y \in W_y, \quad \bar{W}_y \subset U_y, \quad \text{and} \quad \underset{\sim}{1} \notin \bar{W}_y \quad .$$

The family $\{W_y: y \in X - V\}$ is an open covering of the compact set $X - V$, and has a finite subcovering $W_{y_1}, \ldots, W_{y_n}$. For $m = 1, \ldots, n$, let

$$W'_m = f_{y_m}[\bar{W}_{y_m}]$$

and let

$$W' = X' - (W_1' \cup \dots \cup W_n') \quad .$$

Since each W_m' is a closed set not containing $\underset{\sim}{1}'$, we have $\underset{\sim}{1}' \in W' \in \mathcal{S}'$. By Lemma 2.4.7, let $k' \in \mathcal{F}' \cap \mathcal{C}_n \cap \mathcal{C}_{H'}$ be such that

$$k'(\underset{\sim}{1}', \dots, \underset{\sim}{1}') = \underset{\sim}{1}' \text{ and } \breve{k}'[\underset{\sim}{1}'] \subset (W')^n.$$

We now define, for each $z \in X$,

$$a_V(z) = k'(f_{y_1}(z), \dots, f_{y_n}(z)) \quad .$$

It is clear that $a_V \in \mathcal{C}(X, X')$, $a_V(\underset{\sim}{1}) = \underset{\sim}{1}'$, and $a_V(u)H'a_V(v)$ whenever $u H v$. Now, if $a_V(z) = \underset{\sim}{1}'$, then for all m, $f_{y_m}(z) \in W'$ and $z \notin \overline{W_{y_m}}$. Hence $z \notin X - V$.

To continue with the proof of the theorem, suppose that $\underset{\sim}{1} \in V \in \mathcal{S}$. Then the set $V' = X' - a_V[X - V]$ is an open neighborhood of $\underset{\sim}{1}'$ in X'. Let $k' \in \mathcal{K}'$ be such that

$$k'(\underset{\sim}{1}', \underset{\sim}{1}') = \underset{\sim}{1}' \quad \text{and} \quad k'^{\cup}[\underset{\sim}{1}'] \subset (V')^2 \quad .$$

Let $V'' = X' - k'[(X')^2 - (V')^2]$. Define $k_V \in \mathcal{C}_2$ by

$$k_V(x, y) = b_{V''}(k'(a_V(x), a_V(y))) \quad .$$

We check easily that $k_V(\underset{\sim}{1}, \underset{\sim}{1}) = \underset{\sim}{1}$, $k_V \in \mathcal{C}_H$, and $\breve{k}_V[\underset{\sim}{1}] \subset V^2$. Hence the set

$$\mathcal{K} = \{k_V : \underset{\sim}{1} \in V \in \mathcal{S}\}$$

is a k-set of $\mathcal{L}$ such that $\mathcal{K} \subset \mathcal{C}_H$.

Suppose $Y \in X^*$ and $\underset{\sim}{1} \notin \bar{Y}$. Let $V = X - \bar{Y}$, we see that $\underset{\sim}{1} \in V \in \mathcal{S}$. Let $Y' = a_V[Y]$. Since a_V is continuous, $\bar{Y}' = a_V[\bar{Y}]$ and hence $\underset{\sim}{1}' \notin \bar{Y}'$. Let $e' \in \mathcal{E}'$ be such that $e'(Y') \neq \underset{\sim}{1}'$, and let $V' = X' - \{e'(Y')\}$. Define $e_Y \colon X^* \to X$ by

$$e_Y(Z) = b_{V'}(e'(a_V[Z])) \quad \text{for all } Z \in X^* \quad .$$

It is easily checked that $e_Y(Y) \neq \underset{\sim}{1}$, $e_Y(Z) = \underset{\sim}{1}$ whenever $\underset{\sim}{1} \in \bar{Z}$, and e_Y is $\check{H}$-preserving. To prove that $e \in \mathcal{Q}_{\check{H}}$ we have to show that the mapping $a^* \colon X^* \to X'^*$ defined by

$$a^*(Z) = a_V[Z] \quad \text{for all } Z \in Z^*$$

is continuous. Let U^* be an element of the subbase for the topology $(\mathcal{S}')^*$ on $(X')^*$. Then there exists $U' \in \mathcal{S}'$ such that either

$$(1) \qquad U^* = \{Z' \in X'^*: \ \overline{Z'} \subset U'\}$$

or else

$$(2) \qquad U^* = \{Z' \in X'^*: \ Z' \cap U' \neq 0\} \quad .$$

If (1) holds, then

$$(a^*)^{\smile}[U^*] = \{Z \in X^*: \ \bar{Z} \subset \breve{a}_V[U']\} \in \mathcal{S}^* \quad ,$$

and if (2) holds, then

$$(a^*)^{\smile}[U^*] = \{Z \in X^*: Z \cap \breve{a}_V[U'] \neq 0\} \in \mathcal{S}^* \quad .$$

Since every open set of $(X')^*$ is a union of finite intersections of sets U^* of the form (1) or (2), it follows that $(a^*)^{\smile}[W]$ is open in X^* whenever $W \in \mathcal{S}'^*$. Hence a^* is continuous. We conclude that the set

$$\mathcal{E} = \{e_Y: \ Y \in X^* \ \text{and} \ 1 \notin \bar{Y}\}$$

is an e-set of $\mathcal{L}$ such that $\mathcal{E} \subset \mathcal{Q}^{\smile}_H$. Our proof is complete.

If we examine the proof of Theorem 2.6.4 carefully, we see that at no place have we used the full fact that H orders $(X, \underset{\sim}{0}, \underset{\sim}{1})$ and H' orders $(X', \underset{\sim}{0}', \underset{\sim}{1}')$. This observation leads to the following

COROLLARY 2.6.5. Suppose that $\mathcal{L}$ and $\mathcal{L}'$ are continuous logics such that:

(a) $\mathcal{L}'$ has a t-set, a k-set, and an e-set ;

(b) whenever $x, y \in X$ and $x \neq y$, there is a function $a \in \mathcal{C}(X, X')$ such that $a(x) \neq a(y)$;

(c) whenever $\underset{\sim}{1}' \in V' \in \mathcal{S}'$, there is a function $b \in \mathcal{C}(X', X)$ such that $b(\underset{\sim}{1}') = \underset{\sim}{1}$ and $b^{\smile}[\underset{\sim}{1}] \subset V'$.

Then $\mathcal{L}$ has a t-set, a k-set, and an e-set.

In case $\mathcal{L}'$ is the classical two-valued logic ℓ, Theorem 2.6.4 and Corollary 2.6.5 reduce to parts .2) and .1) respectively of Theorem 2.6.1. On the other hand, if $\mathcal{L}'$ is the classical [0, 1]-valued logic discussed in Example 2.5.2, Theorem 2.6.3 is seen to be an immediate consequence of Corollary 2.6.5. Indeed, (a) is satisfied by this choice of $\mathcal{L}'$, (b) follows from Urysohn's Lemma, and (c) holds because X has a nontrivial arc.

Simple examples show that there exist continuous logics $\mathcal{L}$, even with $\mathcal{F} = \mathcal{C} \cup \mathcal{D}$, which do not have any t-set. For example, if the point $\underset{\sim}{1}$ is isolated in X, or even has a neighborhood basis composed of closed open sets in X, but X has some infinite connected subset, then $\mathcal{L}$ has no t-set. On the other hand, it may be seen from the proof of Theorem 2.6.1 that if $\mathcal{F} = \mathcal{C} \cup \mathcal{D}$ and the point $\underset{\sim}{1}$ is isolated, or has a neighborhood basis composed of closed open sets in X, then $\mathcal{L}$ does have both a k-set and an e-set. It is also easily seen that if X is a compact topological group and $\mathcal{F} = \mathcal{C} \cup \mathcal{D}$, then $\mathcal{L}$ must have a t-set, which can be defined in the obvious way using translations.

Some open problems of a purely topological character are the following.

1. Does every $\mathcal{L}$, for which $\mathcal{F} = \mathcal{C} \cup \mathcal{D}$ and X is a compact topological group, have a k-set and an e-set, as well as a t-set?

2. Is there any $\mathcal{L}$ with $\mathcal{F} = \mathcal{C} \cup \mathcal{D}$ which does not have a k-set and an e-set?

3. Is there a continuous logic $\mathcal{L}$ with $\mathcal{F} = \mathcal{C} \cup \mathcal{D}$ such that none of the logics $\mathcal{L}(\underset{\sim}{0}', \underset{\sim}{1}')$, $\underset{\sim}{0}', \underset{\sim}{1}' \in X$, have a t-set?

4. What are some other (useful) characterizations of continuous logics which have t-sets, k-sets, and e-sets?

CHAPTER III

MODEL-THEORETIC PRELIMINARIES

3.1. Models

We present in this chapter some semantical counterparts to the notions given in Chapter II. We shall introduce mathematical objects, called *models*, which will be the subject of study for the formal languages of Chapter II.

By a *model* A for the continuous logic $\mathcal{L}$ is meant a pair $A = (R, \mathscr{A})$ where

3.1.1. R is a non-empty set ;

3.1.2. $\mathscr{A}$ is a function whose domain is the set of all predicate and constant symbols of $\mathcal{L}$;

3.1.3. $\mathscr{A}(P_\eta)$: $R^{\tau(\eta)} \to X$ for each $\eta < \pi$;

3.1.4. $\mathscr{A}(c_\zeta) \in R$ for each $\zeta < \kappa$.

We speak of R as the set of *elements of* the model A and the elements $\mathscr{A}(c_\zeta)$, $\zeta < \kappa$, as the *distinquished elements of* A. Each function $\mathscr{A}(P_\eta)$, $\eta < \pi$, may be regarded as a partition of the set $R^{\tau(\eta)}$ into (at most) $|X|$ disjoint subsets with each subset giving rise to a point of X and distinct subsets giving rise to distinct points of X. We refer to the pair $< \tau, \kappa >$ as the *similarity type* of the model A and to the space X as the *value space* for A. Thus the similarity type $< \tau, \kappa >$ of a model A for $\mathcal{L}$ coincides with the type of $\mathcal{L}$, and is uniquely determined by A. On the other hand, the value space X for A is not determined by A but by $\mathcal{L}$. Hence, a model A for $\mathcal{L}$ may also be a model for a logic $\mathcal{L}'$ which is different from $\mathcal{L}$; in general, a model for $\mathcal{L}$ is always a model for any logic $\mathcal{L}'$ whose value space X' includes X and whose type is the same as the type of $\mathcal{L}$. Since the logic $\mathcal{L}$ is assumed fixed, whenever we speak of models we shall mean models for $\mathcal{L}$; we let $\mathscr{M}$, or sometimes $\mathscr{M}_{\mathcal{L}}$, denote the class of all models for $\mathcal{L}$, and we shall use the capital Roman letters A, B, and C, possibly with subscripts, to range over elements of $\mathcal{M}$. We shall also always write $A = (R, \mathcal{A})$ and $B = (S, \mathcal{B})$; the script letter $\mathcal{C}$, of course, has already been appropriated for the set of continuous functions, so we shall refrain from writing $C = (T, \mathcal{C})$. We say that A is finite if R is finite and A is infinite if R is infinite. More generally, by the *cardinal of* A, in symbols $\|A\|$, we mean the cardinal $|R|$.

If $X = \{0, 1\}$, then each $\mathcal{A}(P_\eta)$, $\eta < \pi$, gives rise to the $\tau(\eta)$-ary relation $(\mathcal{A}(P_\eta))^{\cup}[1]$ over R which determines $\mathcal{A}(P_\eta)$ uniquely. Thus, in the case of the classical two-valued logic ℓ, each model A for ℓ becomes a model in the usual sense.

The model A is a *submodel* of the model B, in symbols $A \subset B$ (due to our consistent use of A, B for models, it will always be clear from context that this is not the subset relation), if

3.1.5. $R \subset S$;

3.1.6. $\mathcal{A}(P_\eta) = \mathcal{B}(P_\eta) \upharpoonright R^{\tau(\eta)}$ for each $\eta < \pi$;

3.1.7. $\mathcal{A}(c_\zeta) = \mathcal{B}(c_\zeta)$ for each $\zeta < \kappa$.

If A is a submodel of B, we say that B is an *extension* of A. Clearly, the relation of being a submodel is reflexive, transitive, and anti-symmetric. Furthermore, each non-empty subset R of S containing the distinguished elements $\mathcal{B}(c_\zeta)$, $\zeta < \kappa$, uniquely determines a submodel A of B satisfying 3.1.5—3.1.7 above. Such a submodel A is said to be *generated by* the subset R. If $A \subset B$, then $\|A\| \leq \|B\|$.

Let ν be an ordinal. Suppose that for each $\xi < \nu$, $A_\xi = (R_\xi, \mathcal{A}_\xi)$ is a model and suppose that whenever $\eta \leq \xi < \nu$, $A_\eta \subset A_\xi$. The *union of the chain of models* A_ξ, $\xi < \nu$, is the model $A = \bigcup_{\xi < \nu} A_\xi$ where

3.1.8. $R = \bigcup_{\xi < \nu} R_\xi$;

3.1.9. $\mathcal{A}(P_\eta) = \bigcup_{\xi < \nu} \mathcal{A}_\xi(P_\eta)$ for each $\eta < \pi$;

3.1.10. $\mathcal{A}(c_\zeta) = \mathcal{A}_\xi(c_\zeta)$ for each $\zeta < \kappa$ and $\xi < \nu$.

Notice that there is no ambiguity in the definition of $\mathcal{A}(c_\zeta)$, $\zeta < \kappa$, and that each $\mathcal{A}(P_\eta)$: $R^{\tau(\eta)} \to X$. Thus $A \in \mathcal{M}$ and $A_\xi \subset A$ for each $\xi < \nu$.

Two models A and B are *isomorphic*, in symbols $A \cong B$, if there exists a one-to-one mapping h such that

3.1.11. $\mathcal{D}h = R$ and $\mathcal{R}h = S$;

3.1.12. $\mathcal{A}(P_\eta)(a_1, \ldots, a_{\tau(\eta)}) = \mathcal{B}(P_\eta)(ha_1, \ldots, ha_{\tau(\eta)})$ for each $\eta < \pi$ and $a_1, \ldots, a_{\tau(\eta)} \in R$;

3.1.13. $h(\mathcal{A}(c_\zeta)) = \mathcal{B}(c_\zeta)$ for each $\zeta < \kappa$.

The function h is called an *isomorphism of* A *onto* B and B is an *isomorphic image of* A *under* h. The relation $\cong$ is an equivalence relation over $\mathcal{M}$, and we speak of the equivalence classes of $\mathcal{M}$ as *isomorphism types* of $\mathcal{M}$. A necessary condition for $A \cong B$ is that $\|A\| = \|B\|$.

The model A is *isomorphically embeddable in* B if A is isomorphic to some submodel of B (or, equivalently, B is isomorphic to to some extension of A). The isomorphic embeddability relation is reflexive and transitive but not anti-symmetric, and it contains the submodel relation as a subrelation.

A more general notion than an isomorphism of A onto B is the following notion of a *homomorphism of* A *onto* B. Two models A and B are *homomorphic* (with respect to $\mathcal{L}$), in symbols A H B, if there exists a mapping h such that

3.1.14. $(\mathcal{A}(P_\eta)(a_1, \ldots, a_{\tau(\eta)}))\ H(\mathcal{B}(P_\eta)(ha_1, \ldots, ha_{\tau(\eta)}))$ for each $\eta < \pi$ and $a_1, \ldots, a_{\tau(\eta)} \in R$;

3.1.16. $h(\mathcal{A}(c_\zeta)) = \mathcal{B}(c_\zeta)$ for each $\zeta < \kappa$.

We shall say that h is a *homomorphism of* A *onto* B and that B is a *homomorphic image of* A *under* h. If $A \cong B$, then A H B. If A H B, then $\|B\| \le \|A\|$; if in addition B H C, then A H C. In case H is the identity relation over X, any one-to-one homomorphism of A onto B is an isomorphism of A onto B.

For the moment we shall not introduce any more algebraic operations on models.

It has already been pointed out that for the logic ℓ the notion of a model for ℓ, i.e., the class $\mathcal{M}_\ell$, coincides with the usual notion of a model. It is easy to see that each of the notions we have introduced reduces, in the case of the logic ℓ, to the corresponding two-valued notion for ℓ.

Let μ be an ordinal. We let $\mathcal{M}_{\mathcal{L}(\mu)}$ denote the class of all models for $\mathcal{L}(\mu)$. We shall often wish to extend a model A in $\mathcal{M}$ to a model in $\mathcal{M}_{\mathcal{L}(\mu)}$ in the following way. Suppose that $a \in R^{\mu}$. We shall denote by (A, a), or sometimes by $(A, a_\zeta)_{\zeta < \mu}$, that model $(R, \mathcal{A}')$ in $\mathcal{M}_{\mathcal{L}(\mu)}$ such that

$$\mathcal{A} \subset \mathcal{A}'$$

and

$$\mathcal{A}'(c_{\kappa+\zeta}) = a_\zeta \quad \text{for} \quad \zeta < \mu \quad .$$

Obviously each model A' in $\mathcal{M}_{\mathcal{L}(\mu)}$ can be written in a unique way as (A, a), where A is in $\mathcal{M}$. The model A in $\mathcal{M}$ is called the *reduct of* A'.

We mention in passing that the simple algebraic notions we have introduced for $\mathcal{M}$ can be extended in an obvious way to models (A, a) in $\mathcal{M}_{\mathcal{L}(\mu)}$. For instance, it would follow from the extended definitions that if $(A, a) \subset (B, b)$ then $A \subset B$. Similarly for $\cong$ and H.

We conclude the introduction of models with a simple lemma.

LEMMA 3.1.17. Let μ be an ordinal, α be a cardinal, and β be the cardinal of the set of isomorphism types of models $(A, a) \in \mathcal{M}_{\mathcal{L}(\mu)}$ where $\|A\| \leq \alpha$. Then

$$\beta \leq \omega \cup |X|^{\alpha} \cup |X|^{|\pi|} \cup \alpha^{|\kappa+\mu|} \quad .$$

PROOF. Let R be a set of cardinality α and suppose that α is infinite. Clearly, each $B' \in \mathcal{M}_{\mathcal{L}(\mu)}$ with $\|B'\| \leq \alpha$ is isomorphically embeddable in some model $A' \in \mathcal{M}_{\mathcal{L}(\mu)}$ with R as its set of elements. Therefore, an upper bound for β is given by the number of models A" in $\mathcal{M}_{\mathcal{L}(\mu)}$ which have a subset of R as its set of elements. There are 2^{α} ways to choose the set R" of elements of $\mathcal{A}''$; since $|X| \geq 2$, we have $2^{\alpha} \leq |X|^{\alpha}$. For each $\eta < \pi$, the number of mappings on $R''^{\tau(\eta)}$ into X is at most $|X|^{\alpha}$ (remembering that each $\tau(\eta)$ is finite). Hence the number of ways to choose all of the functions $\mathcal{A}''(P_\eta)$, $\eta < \pi$, is at most $(|X|^{\alpha})^{|\pi|} = |X|^{\alpha} \cup |X|^{|\pi|}$. The number of ways to choose the distinguished elements $\mathcal{A}''(c_\zeta)$, $\zeta < \kappa + \mu$, of A" is at most $\alpha^{|\kappa+\mu|}$. Hence

$$\beta \leq |X|^{\alpha} \cup |X|^{|\pi|} \cup \alpha^{|\kappa+\mu|} .$$

If α is finite and one of X, π, $\kappa+\mu$ is infinite then the same bound for β will hold. If each of α, X, π, $\kappa+\mu$ is finite then ω is an upper bound for β. The lemma is proved.

3.2. Truth values

We recall that R^{∞} is the set of all eventually constant functions in R^{ω}. If $|R| \geq 2$, then $|R^{\infty}| = \omega \cup |R|$. Elements of R^{∞}, S^{∞} are denoted by small Roman letters r, s, respectively.

Let $\varphi \in \Phi$, $A \in \mathscr{M}$, and $r \in R^{\infty}$. We shall define an element of X, denoted by $\varphi[A, r]$, by recursion on the formulas $\varphi \in \Phi$. We call the element $\varphi[A, r]$ of X the *truth value of* φ *for* the pair A, r. First we define the *value of* a *term* (that is, a variable or a constant of $\mathscr{L}$) for the pair A, r, as follows.

3.2.1. $v_n[A, r] = r_n$, for each $n < \omega$.

3.2.2. $c_{\zeta}[A, r] = \mathscr{A}(c_{\zeta})$, for each $\zeta < \kappa$.

In the next two definitions we define the truth value for atomic formulas.

3.2.3. If x_1, x_2 are terms, then

$$(x_1 \equiv x_2)[A, r] = \underset{\sim}{1} \text{ iff } x_1[A, r] = x_2[A, r] \ ;$$

$$(x_1 \equiv x_2)[A, r] = \underset{\sim}{0} \text{ iff } x_1[A, r] \neq x_2[A, r] \ .$$

3.2.4. If $\eta < \pi$ and $x_1, \ldots, x_{\tau(\eta)}$ are terms, then

$$P_{\eta}(x_1, \ldots, x_{\tau(\eta)})[A, r] = \mathscr{A}(P_{\eta})(x_1[A, r], \ldots, x_{\tau(\eta)}[A, r]) \ .$$

We now extend the definition to arbitrary formulas.

3.2.5. If $f \in \mathscr{C}_n \cap \mathscr{F}$ and $\varphi_1, \ldots, \varphi_n \in \Phi$, then

$$f(\varphi_1, \ldots, \varphi_n)[A, r] = f(\varphi_1[A, r], \ldots, \varphi_n[A, r]) \ .$$

3.2.6. If $q \in \mathscr{Q} \cap \mathscr{F}$, $\varphi \in \Phi$, and $m < \omega$, then

$$(qv_m)\varphi[A, r] = q(\{\varphi[A, r'] : r' \in R^{\infty} \text{ and } r'_n = r_n \text{ for all } n \neq m\}) \ .$$

When necessary, we write $\varphi[A, r]_{\mathscr{L}}$ for $\varphi[A, r]$. It should be clear that by 3.2.1—3.2.6 we have assigned to each formula $\varphi \in \Phi$, model $A \in \mathcal{M}$, and sequence $r \in R^\infty$, a unique point $\varphi[A, r]$ of X.

LEMMA 3.2.7. Let $\varphi \in \Phi$, and $r, r' \in R^\infty$. If $v_n[A, r] = v_n[A, r']$ for all the free variables v_n of φ, then $\varphi[A, r] = \varphi[A, r']$.

PROOF. By an easy induction on the formulas $\varphi \in \Phi$.

It follows from the above lemma that if $\varphi \in \Sigma$, i.e., φ is a sentence, then the truth value $\varphi[A, r]$ is independent of r. For $\varphi \in \Sigma$ the truth value $\varphi[A, r]$ is denoted by $\varphi[A]$, or sometimes by $\varphi[A]_{\mathcal{L}}$.

Turning briefly to the two-valued logic ℓ and the sets Φ_ℓ, Σ_ℓ, $\mathcal{M}_\ell$, we say that a formula $\varphi \in \Phi_\ell$ is *satisfied in* the model $A \in \mathcal{M}_\ell$ *by the sequence* $r \in R^\infty$ if $\varphi[A, r] = \underset{\sim}{1}$. A sentence $\varphi \in \Sigma_\ell$ *holds in* A if $\varphi[A] = \underset{\sim}{1}$, in which case A is said to be a *model of* φ. If $\varphi_1, \varphi_2 \in \Sigma_\ell$ and $A \in \mathcal{M}_\ell$, then

$\mathrm{id}(\varphi_1)$ holds in A iff φ_1 holds in A ;

$\neg(\varphi_1)$ holds in A iff φ_1 fails to hold in A ;

$\&(\varphi_1, \varphi_2)$ holds in A iff both φ_1 and φ_2 hold in A ;

$\wp(\varphi_1, \varphi_2)$ holds in A iff either φ_1 or φ_2 holds in A ;

$(\exists v_m)\varphi$ holds in A iff there exist $a \in R$, $r \in R^\infty$,

such that $r_m = a$ and φ is satisfied in A by r.

We see that 3.2.1—3.2.6 give us precisely the classical notions of satisfiability and validity for the language ℓ.

EXERCISE 3A. Suppose $\mathcal{L}$ has a t-set $\mathcal{T}$. Then the following are equivalent:

(i) For each $\varphi \in \Sigma$, $\varphi[A] = \varphi[B]$.

(ii) For each $\varphi \in \Sigma$, if $\varphi[A] = \underset{\sim}{1}$ then $\varphi[B] = \underset{\sim}{1}$.

EXERCISE 3B. Suppose $\mathcal{L}$ has an H-set $\mathcal{H}$. Then the following are equivalent:

(i) For each $\varphi \in \Sigma$, $\varphi[A] \; H \; \varphi[B]$.

(ii) For each $\varphi \in \Sigma$, if $\varphi[A] = \underset{\sim}{1}$ then $\varphi[B] = \underset{\sim}{1}$.

With respect to the notions of submodels, extensions, isomorphism, and homomorphism, the truth values behave as follows.

EXERCISE 3C. Suppose that $A \subset B$ and $r \in R^{\infty}$. Then for any formula

$$\varphi \in ((\mathcal{Q}_{\breve{H}} \cup \mathcal{C}_{H}) \cap \mathcal{F})(\mathcal{C} \cap \mathcal{F})\wedge ,$$

we have $\varphi[A, r]\ H\ \varphi[B, r]$. In particular, for any sentence

$$\varphi \in \Sigma \cap ((\mathcal{Q}_{\breve{H}} \cup \mathcal{C}_{H}) \cap \mathcal{F})(\mathcal{C} \cap \mathcal{F})\wedge ,$$

we have $\varphi[A]\ H\ \varphi[B]$.

EXERCISE 3D. Suppose that $A \subset B$ and $r \in R^{\infty}$. Then for any formula

$$\varphi \in ((\mathcal{Q}_{H} \cup \mathcal{C}_{H}) \cap \mathcal{F})(\mathcal{C} \cap \mathcal{F})\wedge ,$$

we have $\varphi[B, r]\ H\ \varphi[A, r]$. In particular, for any sentence

$$\varphi \in \Sigma \cap ((\mathcal{Q}_{H} \cup \mathcal{C}_{H}) \cap \mathcal{F})(\mathcal{C} \cap \mathcal{F})\wedge ,$$

we have $\varphi[B]\ H\ \varphi[A]$.

EXERCISE 3E. Suppose that $A \cong B$ under the isomorphism h, $r \in R^{\infty}$, $s \in S^{\infty}$, and $s = h \circ r$. Then for any formula $\varphi \in \Phi$, we have $\varphi[A, r] = \varphi[B, s]$. In particular, for any sentence $\varphi \in \Sigma$, $\varphi[A] = \varphi[B]$.

EXERCISE 3F. Suppose that $\underset{\sim}{0}\ H\ \underset{\sim}{1}$. Let

$$\mathcal{G} = \mathcal{F} \cap (\mathcal{C}_{H} \cup \mathcal{Q}_{H*}) .$$

Suppose that $A\ H\ B$ under the homomorphism h, $r \in R^{\infty}$, $s \in S^{\infty}$, and $s = h \circ r$. Then for any formula $\varphi \in \mathcal{G}(\wedge) \subset \Phi$, we have $\varphi[A, r]\ H\ \varphi[B, s]$. In particular, for any sentence $\varphi \in \Sigma \cap \mathcal{G}(\wedge)$, $\varphi[A]\ H\ \varphi[B]$.

The following exercises deal with the effect of varying the logic $\mathcal{L}$ on the truth value of a formula φ.

EXERCISE 3G. Suppose $\mathcal{F}' \subset \mathcal{C} \cup \mathcal{Q}$, $\varphi \in \Phi_{\mathcal{L}} \cap \Phi_{\mathcal{L}(\mathcal{F}')}$, $A \in \mathcal{M}$, and $r \in R^{\infty}$. Then

$$\varphi[A, r]_{\mathcal{L}} = \varphi[A, r]_{\mathcal{L}(\mathcal{F}')} .$$

Notice that, on the other hand, a change in $\underset{\sim}{0}$ or $\underset{\sim}{1}$ will cause a change in the truth values of formulas such as $v_0 \equiv v_1$.

EXERCISE 3H. Let μ be an ordinal, $\varphi \in \Phi$, $a \in R^{\mu}$, and $r \in R^{\infty}$. Then we have

$$\varphi[(A, a), r]_{\mathscr{L}(\mu)} = \varphi[A, r]_{\mathscr{L}} .$$

Hereafter we may write $\varphi[A, r]$ for either $\varphi[A, r]_{\mathscr{L}}$ or $\varphi[A, r]_{\mathscr{L}(\mathscr{F}')}$, and we may write $\varphi[(A, a), r]$ for $\varphi[(A, a), r]_{\mathscr{L}(\mu)}$.

EXERCISE 3I. Let μ be an ordinal. Let a_{ζ}, $\zeta < \mu$, and b_{ζ}, $\zeta < \mu$, be enumerations of A and B, respectively. If $\varphi[(A, a)] = \varphi[(B, b)]$ for each atomic sentence φ of $\mathscr{L}(\mu)$, then $A \cong B$. If $\varphi[(A, a)]\ H\ \varphi[(B, b)]$ for each atomic sentence φ of $\mathscr{L}(\mu)$, then $A\ H\ B$.

Let $\varphi \in \Phi$ and let $c_{\zeta_0}, \ldots, c_{\zeta_n}$ be constant symbols of $\mathscr{L}$. We shall denote by $\varphi(c_{\zeta_0}, \ldots, c_{\zeta_n})$ the formula obtained from φ by replacing all free occurences of the variables $v_0, \ldots, v_n$ in φ by c_{ζ_0}, $\ldots$, c_{ζ_n}, respectively. The following three exercises contain all that we shall need to know about the above notion of substitution. They will be used in some later proofs.

EXERCISE 3J. Give a precise recursive definition of $\varphi(c_{\zeta_0}, \ldots, c_{\zeta_n})$.

EXERCISE 3K. If $\varphi \in \Sigma_{\mathscr{L}(\mu)}$, then there exist $n < \omega$, $\psi \in \Phi$, and $\zeta_0, \ldots, \zeta_n < \kappa + \mu$ such that $\varphi = \psi(c_{\zeta_0}, \ldots, c_{\zeta_n})$.

EXERCISE 3L. Show that if all the free variables of φ are among $v_0, \ldots, v_n$, then $\varphi(c_{\zeta_0}, \ldots, c_{\zeta_n})$ is a sentence.

EXERCISE 3M. Suppose that $r \in R^{\infty}$ and for each i, $0 \leq i \leq n$, $r_i = \mathscr{A}(c_{\zeta_i})$. Then

$$\varphi[A, r] = \varphi(c_{\zeta_0}, \ldots, c_{\zeta_n})[A, r] .$$

EXERCISE 3N. Assume that $A \subset B$ and $a \in R$. Then the following are equivalent:

(i) For all $r \in R^\infty$ and all $\varphi \in \Phi$,

$$\varphi[A, r] = \varphi[B, r] \quad .$$

(ii) For all $r \in R^\infty$ and $\psi \in \Phi_{\mathcal{L}(1)}$,

$$\psi[(A, a), r] = \psi[(B, a), r] \; .$$

EXERCISE 30. Assume that $A \subset B$. Then the following are equi-lent:

(i) For all $r \in R^\infty$ and all $\varphi \in \Phi$,

$$\varphi[A, r] = \varphi[B, r] \; .$$

(ii) For all ordinals μ, all $a \in R^\mu$, all $\psi \in \Phi_{\mathcal{L}(\mu)}$, and all $r \in R^\infty$,

$$\psi[(A, a), r] = \psi[(B, a), r] \quad .$$

(iii) There exists an enumeration $a \in R^\nu$ of R such that for all $\psi \in \Sigma_{\mathcal{L}(\nu)}$,

$$\psi[(A, a)] = \psi[(B, a)] \quad .$$

(Hint: From (i) to (ii) use Ex. 3N as the first step of a ansfinite induction. From (iii) to (i) use Ex. 3M.)

3. The elementary topology

We turn our attention to some important notions concerning the uth value function $\varphi[A, r]$. It turns out that many of the things we do this section depend only on the existence of the function $\varphi[A, r]$, and t on the particular manner in which $\varphi[A, r]$ is defined. As usual, how-er, it is best to keep in mind the meaning of $\varphi[A, r]$, especially its aning in the logic ℓ.

Consider the product space X^Σ with the usual product topology . By Tychonoff's theorem, the space $(X^\Sigma, \mathcal{S}^\Sigma)$ is compact and Hausdorff. call from Chapter I that a closed set Y of X^Σ is called singular if r some $\varphi \in \Sigma$ and some closed set Z of X, where $X - Z \in \mathcal{S}_0$, we have

$$Y = \{h \in X^\Sigma : \; h(\varphi) \in Z\} \quad ,$$

d a closed set of X^Σ is called basic if it is a finite union of singular osed sets of X^Σ.

An (*elementary*) *theory* Θ is an element of $S(X^\Sigma)$, that is, a subset of X^Σ. A theory Θ is *open* (or *closed*) if Θ is an open (or closed) subset of X^Σ. The *closure* of a theory Θ is the set $\bar{\Theta} \subset X^\Sigma$. Clearly, non-empty intersections of closed theories are closed theories, and non-empty unions of open theories are open theories. By introducing the space $(X^\Sigma)^*$ together with the open sets $(\mathcal{S}^\Sigma)^*$, we may speak of the non-empty theories as points of the topological space $(X^\Sigma)^*$.

For each model $A \in \mathcal{M}$, we define $[A]$ (or more precisely $[A]_{\mathcal{L}}$) to be the element of X^Σ such that

$$[A](\varphi) = \varphi[A] \quad \text{for each } \varphi \in \Sigma \quad .$$

We shall call $[A]$ the *value function* of A. Thus the function $[\ ]$ which associates A with its value function is a mapping on $\mathcal{M}$ into X^Σ. We define the *theory of* a subset $K \subset \mathcal{M}$ as follows:

$$Th(K) = \{[A]: A \in K\} \quad .$$

Note that $Th(\{A\}) = \{[A]\}$; we shall write $Th(A)$ for $Th(\{A\})$.

For a theory Θ, let the *class of all models of* Θ be defined as:

$$Mod(\Theta) = \{A \in \mathcal{M}: [A] \in \Theta\} \quad .$$

Thus we have introduced the three functions

$$[\ ] : \mathcal{M} \to X^\Sigma \quad ,$$
$$Th : S(\mathcal{M}) \to S(X^\Sigma) \quad ,$$
$$Mod : S(X^\Sigma) \to S(\mathcal{M}) \quad .$$

It may very well be that $Th(\mathcal{M})$ is a proper subset of X^Σ. For example, in the two-valued logic ℓ, the constant function $\Sigma_\ell \times 1$ belongs to $2^{\Sigma_\ell} - Th(\mathcal{M}_\ell)$. Notice that the theory $\Theta = \{\Sigma_\ell \times 1\}$ has the property that $Th(Mod(\Theta)) = 0$.

EXERCISE 3P. Let $K, K_1, K_2 \subset \mathcal{M}$ and let $\Theta, \Theta_1, \Theta_2$ be theories. Then we have the following relationships.

$$Th(K_1 \cup K_2) = Th(K_1) \cup Th(K_2) \quad ;$$

$$Th(K_1 \cap K_2) \subset Th(K_1) \cap Th(K_2) \quad ;$$

$$Mod(\Theta_1 \cup \Theta_2) = Mod(\Theta_1) \cup Mod(\Theta_2) \quad ;$$

$$Mod(\Theta_1 \cap \Theta_2) = Mod(\Theta_1) \cap Mod(\Theta_2) \quad ;$$

$$K \subset Mod(Th(K)) \; ; \qquad \Theta \supset Th(Mod(\Theta)) \; ;$$

$$Th(K) = Th(Mod(Th(K))) \; ; \qquad Mod(\Theta) = Mod(Th(Mod(\Theta))) \; ;$$

$$\Theta \cap Th(K) \neq 0 \text{ if and only if } Mod(\Theta) \cap K \neq 0.$$

The function Mod induces a natural topology on $\mathcal{M}$ which we shall call the *elementary topology* on $\mathcal{M}$. The open (or closed) sets of $\mathcal{M}$ are the images $Mod(\Theta)$ of open (or closed) sets Θ of X^{Σ}. We classify the closed sets K of $\mathcal{M}$ as follows.

3.3.1. K is a *singular elementary class*, in symbols $K \in EC_s$, if K is the image $Mod(\Theta)$ of a singular closed set Θ of X^{Σ}; that is, $K = \{A \in \mathcal{M} : \varphi[A] \in Y\}$ for some $\varphi \in \Sigma$ and closed set $Y \subset X$.

3.3.2. K is a *basic elementary class*, in symbols $K \in EC$, if K is a finite union of singular elementary classes.

3.3.3. K is an *elementary class*, in symbols $K \in EC_{\Delta}$, if K is an intersection of basic elementary classes.

Again, it should be clear that singular elementary classes form a (closed) subbase for the elementary topology, basic elementary classes form a (closed) base for the elementary topology, and the elementary classes are the closed sets of the elementary topology.

Two models $A, B \in \mathcal{M}$ are *elementarily equivalent* if $Th(A) = Th(B)$, or equivalently $[A] = [B]$. A class $K \subset \mathcal{M}$ is *elementarily closed* if whenever $A \in K$ and $Th(A) = Th(B)$, then $B \in K$. The elementary topology for $\mathcal{M}$ is not Hausdorff. It will become Hausdorff under the identification of elementarily equivalent models. Under this identification, there is a homeomorphism between the space of elementarily equivalence classes of $\mathcal{M}$ and the subspace $Th(\mathcal{M})$ of X^{Σ}.

EXERCISE 3Q.

.1) $K \in EC_\Delta$ iff $K = Mod(\overline{Th(K)})$.

.2) K is elementarily closed iff $K = Mod(Th(K))$.

.3) $Mod(\Theta)$ is elementarily closed for every theory Θ .

.4) Any elementary class is elementarily closed.

.5) If K is elementarily closed, then $K \in EC_\Delta$ iff Th(K) is closed in X^Σ .

.6) The elementary topology is compact iff $Th(\mathcal{M})$ is closed in X^Σ .

Let $K \subset \mathcal{M}$. A theory Θ is K-*consistent* if $\Theta \cap Th(K) \neq 0$; if $\Theta \cap Th(K) = 0$, we say that Θ is K-*inconsistent*. If Θ is K-consistent, then any theory which includes Θ is K-consistent, and in particular $\bar{\Theta}$ is K-consistent. A theory Θ is K-*complete* if $\Theta = Th(A)$ for some $A \in K$. If $K = \mathcal{M}$, we say that a theory is *consistent*, *inconsistent*, or *complete*, without mentioning $\mathcal{M}$.

THEOREM 3.3.4. The following are equivalent:

(i) For every theory Θ, if Θ is an accumulation point in $(X^\Sigma)^*$ of K-consistent theories then $\bar{\Theta}$ is K-consistent.

(ii) The set of theories Θ where $\bar{\Theta}$ is K-consistent is a closed set of $(X^\Sigma)^*$.

(iii) Th(K) is closed in (X^Σ).

PROOF. It is obvious that (i) and (ii) are equivalent. Therefore we shall prove that (ii) and (iii) are equivalent. Assume (ii). Let $h \in \overline{Th(K)}$ and let $\Theta = \{h\}$. We show that

(1) Θ is an accumulation point of closed K-consistent theories.

Suppose $\Theta \in U \in (\mathcal{S}^\Sigma)^*$. We may assume that there exist $V_1, \ldots, V_n \in \mathcal{S}^\Sigma$ such that

$$U = \{Y\colon \ \bar{Y} \subset V_1 \cup \ldots \cup V_n \ \text{ and } \ Y \cap V_m \neq 0 \ \text{ for each } \ m \leq n\} \quad ,$$

and so

$$\bar{\Theta} \subset V_1 \cup \ldots \cup V_n \quad \text{and} \quad \Theta \cap V_m \neq 0 \quad \text{for} \quad m \leq n \ .$$

Since Θ is a singleton, this means that $h \in V_1 \cap \ldots \cap V_n \in \mathcal{S}^{\Sigma}$. Hence there exists $A \in K$ such that $[A] \in V_1 \cap \ldots \cap V_n$. Let $\Theta' = Th(A)$. Θ' is a closed K-consistent theory and $\Theta' \in U$. Thus (1) is proved. Using (i), we see that $\Theta = \bar{\Theta}$ is K-consistent, and hence $h \in Th(K)$. Condition (iii) follows.

Assume (iii). Thus $X^{\Sigma} - Th(K)$ is open, and the set

$$\{\Theta: \bar{\Theta} \subset (X^{\Sigma} - Th(K))\}$$

is open and

$$\{\Theta: \bar{\Theta} \cap Th(K) \neq 0\}$$

is closed. So (ii) is proved.

If we specialize the definition of elementary class to the case of the logic ℓ and the class $\mathcal{M}_{\ell}$, we see that again we obtain the usual notion of elementary class. Due to the presence of the connective $\neg$, each singular elementary class K can be represented as follows:

$$K = \{A: \varphi[A] = 1 \quad \text{and} \quad A \in \mathcal{M}_{\ell}\}, \quad \text{for some sentence} \quad \varphi \in \Sigma_{\ell} \quad ;$$

due to the presence of the connective $\vee$, each basic elementary class K is also a singular elementary class. Thus an elementary class $K \in EC_{\Delta}$ has its usual meaning. It turns out that, in general, if the logic $\mathcal{L}$ has a t-set $\mathcal{T}$ and a k-set $\mathcal{K}$, then the observations we have made for $\mathcal{M}_{\ell}$ can essentially be carried out for the elementary classes of $\mathcal{M}$. In particular, Theorem 3.3.5 below states that if $\mathcal{L}$ has a t-set $\mathcal{T}$ and a k-set $\mathcal{K}$ then the subbase for the topology on $\mathcal{M}$ is also a base.

THEOREM 3.3.5. Suppose $\mathcal{L}$ has a t-set $\mathcal{T}$ and a k-set $\mathcal{K}$. Then every elementary class K is an intersection of singular elementary classes.

PROOF. It is sufficient to prove that every $K \in EC$ is an intersection of singular elementary classes. So let $\varphi_1, \ldots, \varphi_n \in \Sigma$, and $X - Y_1, \ldots, X - Y_n \in \mathcal{S}_0$, and suppose that

$$K = \{A \in \mathcal{M}: \varphi_m[A] \in Y_m \quad \text{for some} \quad m \leq n\}$$

Let

$$L = \bigcap\{K' : K \subset K' \text{ and } K' \in EC_s\} \quad .$$

Clearly, $K \subset L$. The theorem will be proved if we show that whenever $A \notin K$, there exists $K' \in EC_s$ such that $K \subset K'$ and $A \notin K'$; for then we would have $K = L$. Let $A \notin K$. This means that

$$\varphi_m[A] \notin Y_m \quad \text{for each } m \leq n \quad .$$

Since Y_m is closed, for each $m \leq n$ we may pick $V_m \in \mathcal{S}_0$ so that

$$\varphi_m[A] \in V_m \subset X - Y_m \quad .$$

By Theorem 2.4.5 there is a finite subset

$$\mathcal{T}_m = \{t_{m1}, \ldots, t_{mi_m}\} \subset \mathcal{T}$$

having the property that :

for all $t \in \mathcal{T}_m$, $t(\varphi_m[A]) = \underset{\sim}{1}$, and

if $t(x) = \underset{\sim}{1}$ for every $t \in \mathcal{T}_m$ then $x \in V_m$.

Since $Y_m \cap V_m = 0$, this means that for each $m \leq n$ the set

$$Z_m = \{< t_{m1}(x), \ldots, t_{mi_m}(x) >: x \in Y_m\}$$

is closed in X^{i_m} and

$$Z_m \subset X^{i_m} - \{< \underset{\sim}{1}, \ldots, \underset{\sim}{1} >\} \quad .$$

For each $m \leq n$, let $\underset{\sim}{1} \in U_m \in \mathcal{S}_0$ be such that

$$Z_m \cap U_m^{i_m} = 0 \quad .$$

Let $U = U_1 \cap \ldots \cap U_n$. We see that $\underset{\sim}{1} \in U$ and

$$Z_m \cap U^{i_m} = 0 \quad \text{for each } m \leq n \quad .$$

Let $p = \Sigma_{m \leq n}\, i_m$. By Lemma 2.4.7 there exists a function k obtained from members of $\mathcal{K}$ by composition such that

$$k \in \mathcal{C}_p,\ k(\underset{\sim}{1}, \ldots, \underset{\sim}{1}) = \underset{\sim}{1} \text{ and } \breve{k}[\underset{\sim}{1}] \subset U^p \quad .$$

Now, $k[X^p - U^p]$ is a closed set of $X - \{\underset{\sim}{1}\}$. So let Z be a basic closed set of X such that

$$\underset{\sim}{1} \notin Z \text{ and } k[X^p - U^p] \subset Z \quad .$$

Let $\varphi \in \Sigma$ be defined by

$$\varphi = k(t_{11}(\varphi_1), \ldots, t_{1i_1}(\varphi_1), \ldots, t_{n1}(\varphi_n), \ldots, t_{ni_n}(\varphi_n))$$

and let $K' \in EC_s$ be defined by

$$K' = \{A \in \mathscr{M}: \varphi[A] \in Z\} \quad .$$

It is already clear that $A \notin K'$ because $\varphi[A] = \underset{\sim}{1} \notin Z$. We show that $K \subset K'$. Suppose $B \in K$; then for some $m \leq n$, $\varphi_m[B] \in Y_m$. Thus

$$< t_{m1}(\varphi_m[B]), \ldots, t_{mi_m}(\varphi_m[B])> \quad \in \quad Z_m \quad ,$$

$$< t_{11}(\varphi_1[B]), \ldots, t_{ni_n}(\varphi_n[B])> \quad \in \quad X^p - U^p \quad ,$$

and

$$k(t_{11}(\varphi_1[B]), \ldots, t_{ni_n}(\varphi_n[B])) \quad \in \quad Z \quad .$$

This means that $\varphi[B] \in Z$, so $B \in K'$. The theorem is proved.

CHAPTER IV

ELEMENTARILY EQUIVALENT MODELS

4.1. The extended theory of a model

Let Σ^+ denote the set of all formulas $\varphi(v_0)$ of $\mathcal{L}$ which have at most the single free variable v_0. We thus have $\Sigma \subset \Sigma^+ \subset \Phi$. Obviously, if η is a non-zero limit ordinal, then $\Sigma^+_{\mathcal{L}(\eta)} = \bigcup_{\mu < \eta} \Sigma^+_{\mathcal{L}(\mu)}$. Consider the product space X^{Σ^+} with the usual product topology $\mathscr{S}^{\Sigma^+}$. The space $(X^{\Sigma^+}, \mathscr{S}^{\Sigma^+})$ is compact and Hausdorff. If $\varphi \in \Sigma^+$, $A \in \mathscr{M}$, and $r \in R^\infty$, then the truth value $\varphi[A, r]$ depends only on r_0 and we may write $\varphi[A, r_0]$ for $\varphi[A, r]$. For each $A \in \mathscr{M}$ and $r_0 \in R$ we let $[A, r_0]$ be the element of X^{Σ^+} such that

$$[A, r_0](\varphi) = \varphi[A, r_0] \quad \text{for each} \quad \varphi \in \Sigma^+ \quad .$$

For $A \in \mathscr{M}$, we define the *extended theory* of A to be the set

$$\mathrm{Th}^+(A) = \{[A, r_0]: \ r_0 \in R\} \quad .$$

We are interested in the connection between $\mathrm{Th}(A)$ and $\mathrm{Th}^+(A)$. It is easily seen that if $A \cong B$, then $\mathrm{Th}^+(A) = \mathrm{Th}^+(B)$. On the other hand, if A and B are elementarily equivalent, i.e., $\mathrm{Th}(A) = \mathrm{Th}(B)$, what can we say about $\mathrm{Th}^+(A)$ and $\mathrm{Th}^+(B)$? We provide an answer in Theorem 4.1.5 below in case $\mathcal{L}$ has a t-set, a k-set, and an e-set.

For each $\Psi \subset \Sigma^+$, the space X^Ψ with the usual topology $\mathscr{S}^\Psi$ is compact and Hausdorff. The natural restriction mapping $\restriction\Psi$ takes each function $h \in X^{\Sigma^+}$ into a function $h \restriction \Psi \in X^\Psi$. For convenience we let

$$\mathrm{Th}^+(A) \restriction \Psi = \{[A, r_0] \restriction \Psi: \ r_0 \in R\} \quad .$$

Using the notation we have introduced we see that for $A \in \mathscr{M}$,

$$\mathrm{Th}^+(A) \restriction \Sigma = \mathrm{Th}(A) \quad .$$

Hence if $\mathrm{Th}^+(A) = \mathrm{Th}^+(B)$, then $\mathrm{Th}(A) = \mathrm{Th}(B)$.

For the remainder of this section we assume that $\mathcal{L}$ has a t-set $\mathcal{T}$, a k-set $\mathcal{K}$, and an e-set $\mathcal{E}$. We let Ψ be an arbitrary subset of Σ^+. For each $h \in X^\Psi$, we define the set $\Psi_h \subset \Sigma^+$ as follows:

$$\Psi_h = \{t(\varphi): \ \varphi \in \Psi, \ \ t \in \mathcal{T}, \ \text{ and } \ t(h(\varphi)) = \underset{\sim}{1}\} \quad .$$

The symbols $\underset{\sim}{0}$ and $\underset{\sim}{1}$ are also used to denote constant functions whose ranges are $\{\underset{\sim}{0}\}$ and $\{\underset{\sim}{1}\}$ respectively, where the domain is given by the context.

LEMMA 4.1.1. If $h \in \overline{\mathrm{Th}^+(A) \restriction \Psi}$, then $\underset{\sim}{1} \in \overline{\mathrm{Th}^+(A) \restriction \Psi_h}$.

PROOF: It is sufficient to prove that for any $\varphi_1, \ldots, \varphi_n \in \Psi_h$, and $\underset{\sim}{1} \in V \in \mathcal{S}_0$, there exists an $r_0 \in R$ such that

(1) $$\varphi_m[A, r_0] \in V \ \text{ for each } \ m \leq n \quad .$$

We note that for each $m \leq n$, $\varphi_m = t_m(\psi_m)$ where $t_m(h(\psi_m)) = \underset{\sim}{1}$, $t_m \in \mathcal{T}$ and $\psi_m \in \Psi$. Since each t_m is continuous, we have

$$\breve{t}_m[V] \in \mathcal{S} \ \text{ and } \ h(\psi_m) \in \breve{t}_m[V] \ \text{ for each } \ m \leq n \quad .$$

By hypothesis, there exists a $g \in \mathrm{Th}^+(A) \restriction \Psi$ and $r_0 \in R$ such that $g(\psi_m) \in \breve{t}_m[V]$ and $g(\psi_m) = \psi_m[A, r_0]$ for each $m \leq n$. Then $\psi_m[A, r_0] \in \breve{t}_m[V]$, and (1) follows readily.

LEMMA 4.1.2. Suppose that $\Phi_1 \subset \Sigma^+$ and $\underset{\sim}{1} \in \overline{\mathrm{Th}^+(A) \restriction \Phi_1}$. Let $k \in \mathcal{K}$ and let

$$\Phi_2 = \Phi_1 \cup \{k(\psi, \psi'): \ \psi, \psi' \in \Phi_1\} \quad .$$

Then $\underset{\sim}{1} \in \overline{\mathrm{Th}^+(A) \restriction \Phi_2}$.

PROOF. It is sufficient to prove that for any $\varphi_1, \ldots, \varphi_n \in \Phi_2$, and $\underset{\sim}{1} \in V \in \mathcal{S}_0$, there exists an $r_0 \in R$ such that

(1) $$\varphi_m[A, r_0] \in V \ \text{ for each } \ m \leq n \quad .$$

We note that for each $m \leq n$, either $\varphi_m \in \Phi_1$ or there exist ψ_m and ψ'_m in Φ_1 with $\varphi_m = k(\psi_m, \psi'_m)$. Since k is continuous and $k(\underset{\sim}{1}, \underset{\sim}{1}) = \underset{\sim}{1}$, there exists $\underset{\sim}{1} \in U \in \mathcal{S}_0$ such that $k[U^2] \subset V$. Let $W = U \cap V$. By hypothesis, there exists an $r_0 \in R$ such that for each $m \leq n$,

$$\varphi_m[A, r_0] \in W \text{ if } \varphi_m \in \Phi_1$$

and

$$\psi_m[A, r_0], \psi_m'[A, r_0] \in W \text{ if } \varphi_m = k(\psi_m, \psi_m') \quad .$$

From this (1) follows.

LEMMA 4.1.3. Suppose that $\underset{\sim}{1} \in \overline{Th^+(A) \restriction \Psi}$. Then for each $e \in \mathcal{E}$ and $\varphi \in \Psi$, we have

$$(ev_0)\varphi[A] = \underset{\sim}{1} \quad .$$

PROOF. Let $\varphi \in \Psi$ and $e \in \mathcal{E}$. For each $\underset{\sim}{1} \in U \in \mathcal{S}_0$, there exists $r_0 \in R$ such that $\varphi[A, r_0] \in U$. Let

$$Y = \{\varphi[A, r_0]: r_0 \in R\} \quad .$$

Then $\underset{\sim}{1} \in \bar{Y}$, and hence $e(Y) = \underset{\sim}{1}$. It follows that

$$(ev_0)\varphi[A] = e(Y) = \underset{\sim}{1} \quad .$$

THEOREM 4.1.4. Let $h \in X^{\Psi}$. The following two conditions are equivalent:

(i) $h \in \overline{Th^+(A) \restriction \Psi}$;

(ii) $\varphi[A] = \underset{\sim}{1}$ for each $\varphi \in \Sigma \cap \mathcal{E}(\mathcal{K}(\Psi_n))$.

PROOF. Assume (i). Let $\varphi \in \Sigma \cap \mathcal{E}(\mathcal{K}(\Psi_h))$. Then $\varphi = (ev_0)\psi$ for some $e \in \mathcal{E}$, $\psi \in \mathcal{K}(\Psi_h)$. We recall that for $\psi \in \mathcal{K}(\Psi_h)$ it is necessary and sufficient that there exists a finite sequence $\Phi_1 \subset \Phi_2 \subset \ldots \subset \Phi_n \subset \Sigma^+$ such that $\Phi_1 = \Psi_h$, $\psi \in \Phi_n$, and each Φ_{m+1} is obtained from Φ_m in the manner of Lemma 4.1.2. By 4.1.1 and 4.1.2, $\underset{\sim}{1} \in \overline{Th^+(A) \restriction \Phi_n}$. Then by Lemma 4.1.3 we have $\varphi[A] = \underset{\sim}{1}$.

Assuming it is not the case that (i), we shall show that (ii) also fails. So suppose that $h \notin \overline{Th^+(A) \restriction \Psi}$. This means that there exist $\varphi_1, \ldots, \varphi_n \in \Psi$ and $h(\varphi_1) \in U_1 \in \mathcal{S}_0, \ldots, h(\varphi_n) \in U_n \in \mathcal{S}_0$ such that

(1) $$\{<\varphi_1[A, r_0], \ldots, \varphi_n[A, r_0]>: r_0 \in R\} \subset X^n - (U_1 \times \ldots \times U_n) \quad .$$

By Theorem 2.4.5, for each $m \leq n$ there is a finite subset

$$\mathscr{T}_m = \{t_{m1}, \ldots, t_{mi_m}\} \subset \mathscr{T}$$

such that

$$t(h(\varphi_m)) = \underset{\sim}{1} \quad \text{for each} \quad t \in \mathscr{T}_m \ ,$$

and

(2) $$\text{if} \quad t(x) = \underset{\sim}{1} \quad \text{for all} \quad t \in \mathscr{T}_m, \quad \text{then} \quad x \in U_m \ .$$

Let $p = \Sigma_{n \leq m} i_m$ and define a function $g\colon X^n \to X^p$ by

$$g(x_1, \ldots, x_n) = \langle t_{11}(x_1), \ldots, t_{1i_1}(x_1), t_{21}(x_2), \ldots, t_{ni_n}(x_n) \rangle \ .$$

Let

$$Z = X^n - (U_1 \times \ldots \times U_n) \ .$$

By (2), we have

$$g[Z] \subset X^p - \{\underset{\sim}{1}\}^p \ .$$

Since each function t_{mj}, $1 \leq m \leq n$ and $1 \leq j \leq i_m$, is continuous, g is continuous. Then because Z is compact, $g[Z]$ is closed in X^p. It follows that there exists $V \in \mathscr{S}$ such that $\underset{\sim}{1} \in V$ and

$$g[Z] \subset X^p - V^p \ .$$

By Lemma 2.4.7, there exists $k \in \mathcal{C}_p$, obtained from members of $\mathscr{K}$ by composition, such that

$$k(\underset{\sim}{1}, \ldots, \underset{\sim}{1}) = \underset{\sim}{1} \quad \text{and} \quad \check{k}[\underset{\sim}{1}] \subset V^p \ .$$

Then

$$k[g[Z]] \subset X - \{\underset{\sim}{1}\} \ .$$

Since k is continuous and $g[Z]$ is closed, $k[g[Z]]$ is closed. Recalling that $\mathscr{F}$ is closed under composition and $\mathscr{K} \subset \mathscr{F}$, we have $k \in \mathscr{F}$. Let

$$\psi = k(t_{11}(\varphi_1), \ldots, t_{1i_1}(\varphi_1), t_{21}(\varphi_2), \ldots, t_{ni_n}(\varphi_n)) \ .$$

Then $\psi \in \mathscr{K}(\Psi_h)$ and for all $r_0 \in R$ we have

$$\psi[A, r_0] = k(g(\varphi_1[A, r_0], \ldots, \varphi_n[A, r_0])) \ .$$

Hence by (1),

$$Y = \{\psi[A, r_0]\colon \ r_0 \in R\} \subset k[g[Z]] \ .$$

It follows that $\bar{Y} \subset k[g[Z]]$ and so $\underset{\sim}{1} \notin \bar{Y}$. Since $\mathscr{E}$ is an e-set and $\underset{\sim}{1} \notin \bar{Y}$, there is an $e \in \mathscr{E}$ such that $e(Y) \neq \underset{\sim}{1}$. Hence

$$(ev_0)\ \psi\ [A] \neq \underset{\sim}{1} \quad .$$

Clearly $(ev_0)\psi \in \Sigma \cap \mathscr{E}(\mathscr{K}(\Psi_h))$, and (ii) fails. The theorem is proved.

THEOREM 4.1.5. The following are equivalent.

(i) $Th(A) = Th(B)$.

(ii) $\overline{Th^+(A)} = \overline{Th^+(B)}$.

PROOF. The proof of (i) from (ii) is trivial, so we shall concentrate on the direction (i) to (ii). By the symmetry of the situation it is sufficient to show that $Th^+(A) \subset \overline{Th^+(B)}$. Let $h \in Th^+(A)$. Then $h \in \overline{Th^+(A)}$, so by 4.1.4 with $\Psi = \Sigma^+$,

$$\varphi[A] = \underset{\sim}{1} \quad \text{for every} \quad \varphi \in \Sigma \cap \mathscr{E}(\mathscr{K}(\Psi_h)) \quad .$$

By (i) this means

$$\varphi[B] = \underset{\sim}{1} \quad \text{for every} \quad \varphi \in \Sigma \cap \mathscr{E}(\mathscr{K}(\Psi_h))$$

Using 4.1.4 once more, we have $h \in \overline{Th^+(B)}$. The theorem is proved.

EXERCISE 4A. Prove the following improvements of Theorem 4.1.5:

.1) Define, in the space X^Φ,

$$Th^\infty(A) = \{(\lambda\varphi \in \Phi)\varphi[A, r]: \ r \in R^\infty\} \subset X^\Phi \quad .$$

Prove that if $Th(A) = Th(B)$, then

$$\overline{Th^\infty(A)} = \overline{Th^\infty(B)} \quad .$$

.2) In the space $X^{\Sigma_{\mathscr{L}}(\mu)}$, define

$$Th^\mu(A) = \{(\lambda\varphi \in \Sigma_{\mathscr{L}(\mu)})\,\varphi[(A, a)]: \ a \in R^\mu\} \quad .$$

Prove that if $Th(A) = Th(B)$, then

$$\overline{Th^\mu(A)} = \overline{Th^\mu(B)} \quad .$$

Since the two-valued logic ℓ has a t-set, a k-set, and an e-set, Theorem 4.1.5 will hold for models $A, B \in \mathscr{M}_\ell$. The meaning of it is quite clear. In order to show that $\overline{Th^+(A)} = \overline{Th^+(B)}$, let $r_0 \in R$, $\varphi_1, \ldots, \varphi_n \in \Sigma_\ell^+$, and we have to find $s_0 \in S$ so that $\varphi_m[A, r_0] = \varphi_m[B, s_0]$ for each $m \leq n$. By using $\neg$ at the appropriate places, we can transform the sequence $\varphi_1, \ldots, \varphi_n$ into a sequence $\psi_1, \ldots, \psi_n$ where $\psi_m[A, r_0] = 1$ for each $m \leq n$. Let ψ be the conjunction of $\psi_1, \ldots, \psi_n$ and we have $\psi[A, r_0] = 1$. Apply the quantifier $(\exists v_0)$ to ψ to obtain a sentence φ so that $\varphi[A] = 1$. Hence $\varphi[B] = 1$, and reversing the process we find $s_0 \in S$ so that $\varphi_m[A, r_0] = \varphi_m[B, s_0]$ for each $m \leq n$.

4.2. Elementary extensions.

In the preceeding section we studied one relation between models, namely the relation $\overline{Th^+(A)} = \overline{Th^+(B)}$, which implies that A and B are elementarily equivalent. In this section we shall study another relation between A and B which also implies that A and B are elementarily equivalent, this time the relation of B being an elementary extension of A.

A is said to be an *elementary submodel of* B, and B an *elementary extension of* A, in symbols $A \prec B$, if

4.2.1. $A \subset B$, and

4.2.2. $\varphi[A, r] = \varphi[B, r]$ for all $\varphi \in \Phi$ and all $r \in R^\infty$.

As immediate consequences of the definition it is seen that:

4.2.3. $A \prec A$;

4.2.4. if $A \prec B$ and $B \prec C$, then $A \prec C$;

4.2.5. if $A \prec C$, $B \prec C$ and $A \subset B$, then $A \prec B$;

4.2.6. if $A \prec B$, then $Th(A) = Th(B)$ and $Th^+(A) \subset Th^+(B)$.

Notice that even under the strong hypothesis of $A \prec B$, we can not conclude in general that $\overline{Th^+(A)} = \overline{Th^+(B)}$ without assuming that $\mathcal{L}$ has a t-set $\mathcal{T}$, a k-set $\mathcal{K}$, and an e-set $\mathcal{E}$.

The next theorem is a restatement of Exercise 30 using the newly introduced notation and no proof is required.

THEOREM 4.2.7. If $A \subset B$, then the following are equivalent:

(i) $A \prec B$;

(ii) for every ordinal μ and every $a \in R^{\dot{\mu}}$, $(A, a) \prec (B, a)$;

(iii) for some enumeration $a \in R^\nu$, $[(A, a)] = [(B, a)]$.

A model A is said to be *elementarily embeddable* in B if A is isomorphic to some elementary submodel of B (or, equivalently, B is isomorphic to an elementary extension of A). The elementary embeddability relation is reflexive and transitive but is not anti-symmetric. It is a subrelation of the isomorphic embeddability relation.

THEOREM 4.2.8. If $a \in R^{\mu}$ enumerates R, $b \in S^{\mu}$, and $[(A, a)] = [(B, b)]$, then A is elementarily embeddable in B.

PROOF. The theorem follows easily from a modification of Exercise 3I and from Theorem 4.2.7.

We say that a chain of models A_{ξ}, $\xi < \nu$, is an *elementary chain* if $A_{\eta} \prec A_{\xi}$ whenever $\eta \leq \xi < \nu$.

THEOREM 4.2.9. Let ν be a non-zero ordinal and let A_{ξ}, $\xi < \nu$, be an elementary chain. Then $A_{\eta} \prec \bigcup_{\xi < \nu} A_{\xi}$ for all $\eta < \nu$.

PROOF. Let $A = \bigcup_{\xi < \nu} A_{\xi}$. We have already observed in Section 3.1 that $A_{\eta} \subset A$ for all $\eta < \nu$. By Theorem 4.2.7 it is sufficient to show that for each $\varphi \in \Phi$,

(1) $$\text{for all } \eta < \nu, \quad \text{and all } r \in R_{\eta}^{\infty}, \qquad \varphi[A_{\eta}, r] = \varphi[A, r] .$$

We prove (1) by induction on the formulas of Φ. Let Ψ be the set of all formulas of Φ which satisfy (1). It suffices to prove that $\Phi \subset \Psi$. It follows at once from the definitions involved that $\Lambda \subset \Psi$. Suppose $\eta < \nu$, $r \in R_{\eta}^{\infty}$, and $\varphi = f(\varphi_1, \ldots, \varphi_n)$, where $f \in \mathscr{F} \cap \mathscr{C}_n$, and $\varphi_1, \ldots, \varphi_n \in \Psi$. Then by 3.2.5. and the inductive hypothesis, $\varphi[A_{\eta}, r]$ and $\varphi[A, r]$ are both equal to

$$f(\varphi_1[A, r], \ldots, \varphi_n[A, r]) \quad ,$$

and hence $\varphi[A_{\eta}, r] = \varphi[A, r]$.

Suppose $\eta < \nu$, $r \in R_{\eta}^{\infty}$, and $\varphi = (qv_n)\psi$, where $q \in \mathscr{F} \cap \mathscr{Q}$ and $\psi \in \Psi$. For $\eta \leq \xi < \nu$, let

$$Y_\xi = \{\psi[A_\xi, r']: \ r' \in R_\xi^\infty \ \text{ and } \ r'_m = r_m \ \text{ for } \ m \neq n\} \ ,$$

and let

$$Y = \{\psi[A, r']: \ r' \in R^\infty \ \text{ and } \ r'_m = r_m \ \text{ for } \ m \neq n\} \ .$$

Recalling 3.2.6, we see that

$$\varphi[A, r] = (qv_n) \ \psi \ [A, r] = q(Y) \ ,$$

and

$$\varphi[A_\xi, r] = (qv_n) \ \psi \ [A_\xi, r] = q(Y_\xi)$$

for $\eta \leq \xi < \nu$. Hence, we need only show that

$$q(Y_\eta) = q(Y) \ .$$

It is easily seen that $Y_\xi \subset Y$ whenever $\eta \leq \xi < \nu$, because $\psi \in \Psi$. Also, by hypothesis, we have $A_\xi \prec A_\zeta$ and thus $Y_\xi \subset \dot{Y}_\zeta$ whenever $\eta \leq \xi \leq \zeta < \nu$. Finally, $Y_\eta \neq 0$ because $R_\eta^\infty \neq 0$. Therefore Y_ξ, $\eta \leq \xi < \nu$, is an increasing chain of elements of X^* and

$$\bigcup_{\eta \leq \xi < \nu} Y_\xi \subset Y \ .$$

We now show that

$$\text{(2)} \qquad \bigcup_{\eta \leq \xi < \nu} Y_\xi = Y \ .$$

Suppose $y \in Y$. Then for some $r' \in R^\infty$ with $r'_m = r_m$ for $m \neq n$, we have $\psi[A, r'] = y$. Since $R = \bigcup_{\eta \leq \xi < \nu} R_\xi$, $r'_n \in R_\xi$ for some $\eta \leq \xi < \nu$, and hence $r' \in R_\xi^\infty$. Using again the assumption that $\psi \in \Psi$, we see that $\psi[A_\xi, r'] = y$, and thus $y \in Y_\xi$. This verifies (2). By Theorem 1.3.1 we conclude that in the space X^*,

$$Y \in \overline{\{Y_\xi: \ \eta \leq \xi < \nu\}} \ .$$

Since $A_\eta \prec A_\xi$ for $\eta \leq \xi < \nu$, we see that

$$q(Y_\eta) = q(Y_\xi) \quad \text{for} \quad \eta \leq \xi < \nu \ .$$

Hence

$$\{Y_\xi: \ \eta \leq \xi < \nu\} \subset \breve{q}[q(Y_\eta)] \ .$$

The set $\{q(Y_\eta)\}$ is closed in X, and q is continuous, so the set $\breve{q}[q(Y_\eta)]$ is closed in X^*. It follows that $Y \in \breve{q}[q(Y_\eta)]$. This means

$q(Y) = q(Y_\eta)$ and induction is complete.

EXERCISE 4B. Let A_η, $\eta < \nu$, and B_η, $\eta < \nu$, be two elementary chains of models such that $A_\eta \prec B_\eta$ for all $\eta < \nu$. Then $\bigcup_{\eta<\nu} A_\eta \prec \bigcup_{\eta<\nu} B_\eta$. In particular, if $A_\eta \prec B$ for every $\eta < \nu$, then $\bigcup_{\eta<\nu} A_\eta \prec B$.

EXERCISE 4C. Show that even in the logic ℓ, Theorem 4.2.9. is no longer true if we replace $\prec$ everywhere by the notion of an elementarily equivalent submodel.

EXERCISE 4D. Suppose that every model of a consistent theory Θ has a proper elementary extension. Show that Θ has models of arbitrarily high power in the following sense: for every α there exists $A \in \mathrm{Mod}(\Theta)$ such that $\alpha \le \|A\|$.

COROLLARY 4.2.10. Let A_η, $\eta < \nu$, be an elementary chain and let $A = \bigcup_{\eta<\nu} A_\eta$. Then

$$\mathrm{Th}^+(A) = \bigcup_{\eta<\nu} \mathrm{Th}^+(A_\eta) \quad .$$

If, in addition, $\overline{\mathrm{Th}^+(A_\eta)} = \mathrm{Th}^+(A_\xi)$ whenever $\eta \le \xi < \nu$, then for every $\eta < \nu$,

$$\mathrm{Th}^+(A) = \overline{\mathrm{Th}^+(A)} = \mathrm{Th}^+(A_\eta) \quad .$$

PROOF. By Theorem 4.2.9, each $A_\eta \prec A$. Hence, by 4.2.6, $\mathrm{Th}^+(A_\eta) \subset \mathrm{Th}^+(A)$ and

$$\bigcup_{\eta<\nu} \mathrm{Th}^+(A_\eta) \subset \mathrm{Th}^+(A) \quad .$$

On the other hand, if $g \in \mathrm{Th}^+(A)$, then for some $r_0 \in R$, $g = [A, r_0]$. Since $R = \bigcup_{\eta<\nu} R_\eta$, $r_0 \in R_\eta$ for some $\eta < \nu$. So, by Theorem 4.2.7,

$$[A, r_0] = [A_\eta, r_0] \quad ,$$

and $g \in \mathrm{Th}^+(A_\eta)$. The remaining part of the theorem follows immediately.

4.3. The downward Löwenheim-Skolem theorem

In the last section we were concerned with elementary extensions of models. In particular, we see that if the hypothesis of Exercise 4D is satisfied, then we may construct models of Θ with arbitrarily high powers. We shall now examine the problem of going down in power.

THEOREM 4.3.1. Let A be infinite, let $T \subset R$, and let β be a cardinal such that

$$\|\mathcal{L}\| \cup |T| \leq \beta \leq \|A\| \quad .$$

Then there is a B such that

$$T \subset S, \qquad \|B\| = \beta, \qquad \text{and} \qquad B \prec A.$$

PROOF. We first construct the model B, and then show that it has the desired properties. Suppose that $\|A\| = \alpha$. Let R be well-ordered by $a \in R^{\alpha}$. We shall define a function

$$h: \ R^{\infty} \times \omega \times \Phi \times \mathcal{S}_0 \rightarrow R$$

as follows. Let $r \in R^{\infty}$, $n \in \omega$, $\varphi \in \Phi$, and $U \in \mathcal{S}_0$.

> $h(r, n, \varphi, U)$ shall be the first element $a_{\xi} \in R$ such that there exists $r' \in R^{\infty}$, $r'_m = r_m$ for $m \neq n$, $r'_n = a_{\xi}$, and $\varphi[A, r'] \in U$, if such r' exists; $h(r, n, \varphi, U) = a_0$ if no such r' exists.

Clearly h is a well-defined function. Notice that since $\|\mathcal{L}\| \leq \beta$, we have

$$\omega \cup |\mathcal{S}_0| \cup |\Phi| \leq \beta \quad .$$

Let T_0 be any subset of R such that

$$T \cup \{\mathscr{A}(c_{\zeta}): \ \zeta < \kappa\} \subset T_0 \qquad \text{and} \qquad |T_0| = \beta \quad .$$

We shall define the increasing sequence of subsets

$$T_0 \subset T_1 \subset \ldots \subset T_p \subset \ldots, \quad p < \omega$$

as follows. Assume that T_p has been defined. We define

$$T_{p+1} \ = \ T_p \cup h[T_p^{\infty} \times \omega \times \Phi \times \mathcal{S}_0] \quad .$$

By a simple induction, we have $|T_p| = \beta$ for all $p < \omega$. Let $S = \bigcup_{p<\omega} T_p$ and let B be the submodel generated by S. Clearly $|S| = \beta$ and $\|B\| = \beta$. We complete the proof by showing that for all $\varphi \in \Phi$,

(1) $$\varphi[B, s] = \varphi[A, s] \quad \text{for all} \quad s \in S^{\infty} \quad .$$

Let Ψ be the set of formulas of Φ having property (1). We see that $\Lambda \subset \Psi$. Suppose that $\varphi_1, \ldots, \varphi_n \in \Psi$, $f \in \mathscr{F} \cap \mathcal{C}_n$, $\varphi = f(\varphi_1, \ldots, \varphi_n)$.

It can be easily verified that $\varphi \in \Psi$. Suppose that $\psi \in \Psi$, $q \in \mathscr{F} \cap \mathscr{Q}$, and $\varphi = (qv_n)\psi$. Let $s \in S^\infty$,

$$Y_A = \{\psi[A, r']: \quad r' \in R^\infty \quad \text{and} \quad r'_m = s_m \quad \text{for} \quad m \neq n\}$$

and let

$$Y_B = \{\psi[B, s']: \quad s' \in S^\infty \quad \text{and} \quad s'_m = s_m \quad \text{for} \quad m \neq n\} \quad .$$

Since $\psi \in \Psi$ and $S \subset R$, we have

$$Y_B \subset Y_A \quad .$$

We know that $\varphi[A, s] = q(Y_A)$ and $\varphi[B, s] = q(Y_B)$. Hence, in order to complete the induction, we shall prove that $q(Y_A) = q(Y_B)$. Since q is continuous, and since $Y_B \subset Y_A$, it suffices to show that

$$Y_A \subset \bar{Y}_B \quad .$$

Let $y \in Y_A$ and $y \in U \in \mathscr{S}_0$. This means that there exists

$$r' \in R^\infty, \quad r'_m = s_m \quad \text{for} \quad m \neq n, \quad \text{and} \quad \psi[A, r'] \in U \quad .$$

Since $s \in S^\infty$ and its range is finite, for some $p < \omega$, $s \in T_p^\infty$. Hence, by the definition of T_{p+1},

$$h(s, n, \psi, U) \in T_{p+1} \quad .$$

Let $s'_m = s_m$ for $m \neq n$, and $s'_n = h(s, n, \varphi, U)$. Clearly, $s' \in T_{p+1}^\infty \subset S^\infty$, $\psi[A, s'] \in U$, and $\psi[A, s'] \in Y_B$. So

$$Y_B \cap U \neq 0 \quad .$$

Thus, for every $U \in \mathscr{S}_0$ such that $y \in U$, $Y_B \cap U \neq 0$. This implies that $Y_A \subset \bar{Y}_B$, as was to be proved. Now (1) is proved and $B \prec A$.

COROLLARY 4.3.2. Suppose the theory Θ has a model A of power α and $\|\mathscr{L}\| \leq \beta \leq \alpha$. Then Θ has a model of power β.

EXERCISE 4E. The model B constructed in the proof of Theorem 4.3.1 can be modified so as to have the following additional properties:

$$\overline{Th^+(A)} = \overline{Th^+(B)} \quad \text{and}$$

$$\overline{Th^+((A, b))} = \overline{Th^+((B, b))} \quad \text{for all ordinals } \mu \text{ and all } b \in S^\mu .$$

Hint: Define a function

$$g: \ R^\infty \times \omega \times \mathcal{S}_0^\Phi \to R$$

as follows. For $r \in R^\infty$, $n < \omega$, basic open set $V \in \mathcal{S}_0^\Phi$ given by formulas $\varphi_1, \ldots, \varphi_k \in \Phi$, and open sets $U_1, \ldots, U_k \in \mathcal{S}_0$,

> $g(r, n, V)$ is the least element $a_\xi \in R$ for which there exists $r' \in R^\infty$, $r'_m = r_m$ for $m \neq n$, $r'_n = a_\xi$ and such that
>
> $\varphi_i[A, r'] \in U_i$ for each i, $1 \leq i \leq k$;
>
> $g(r, n, V) = a_0$ if no such r' exists.

Now, generate the T_p's using the function g instead of h.

EXERCISE 4F. One can even show that the model B of Theorem 4.3.1 can be constructed so that

$$\overline{Th^\infty((A, b))} = \overline{Th^\infty((B, b))} \text{ for all ordinals } \mu \text{ and } b \in S^\mu .$$

(For the definition of Th^∞, see Exercise 4A.)

EXERCISE 4G. Theorem 4.3.1 remains true even if X is not compact. A similar remark applies to Exercises 4E and 4F. Is the assumption of compactness of X necessary in the proof of Theorem 4.2.9?

EXERCISE 4H. If every model of a consistent theory Θ has a proper elementary extension, then for every cardinal α such that $\|\mathcal{L}\| \leq \alpha$, there exists a model of Θ having power α.

EXERCISE 4I. Suppose $\|\mathcal{L}\| < \alpha$. Then every model A of power α is the union of some elementary chain A_ξ, $\xi < \nu$, such that $\|A_\xi\| < \alpha$ for each $\xi < \nu$.

CHAPTER V

ULTRAPRODUCTS OF MODELS AND APPLICATIONS

5.1. The fundamental lemma

Throughout this section we shall let I be a non-empty set, D be an ultrafilter over the set I, and $F = \lambda i A_i$ be a function on I into $\mathscr{M}$. In Section 1.2 we defined the D-product, D-prod $\lambda i\ R_i$, of the sets R_i, $i \in I$. We shall now extend this definition to models, and define the D-product, D-prod $\lambda i\ A_i$, of the models A_i, $i \in I$. We shall establish the fundamental relation

$$\varphi[\text{D-prod } \lambda i\ A_i] = \text{D-lim } \lambda i\ \varphi[A_i] \text{ for every } \varphi \in \Sigma.$$

Let us recall the following definitions from Section 1.2; let f, g range over $\Pi_{i \in I} R_i$.

$$f \sim g \text{ iff } \{i \in I : f(i) = g(i)\} \in D.$$

$$f^{\sim} = \{g : g \sim f\}.$$

$$\text{D-prod } \lambda i\ R_i = \{f^{\sim} : f \in \Pi_{i \in I} R_i\}.$$

The *D-product*, or *ultraproduct* of the models A_i, $i \in I$, in symbols D-prod $\lambda i\ A_i$, is the model $A = (R, \mathscr{A}) \in \mathscr{M}$ such that:

5.1.1. $R = \text{D-prod } \lambda i\ R_i$;

5.1.2. $\mathscr{A}(c_\zeta) = (\lambda i \mathscr{A}_i(c_\zeta))^{\sim}$, for each $\zeta < \kappa$;

5.1.3. For each $\eta < \pi$ and $f_1^{\sim}, \ldots, f_{\tau(\eta)}^{\sim} \in R$,

$$A(P_\eta)(f_1^{\sim}, \ldots, f_{\tau(\eta)}^{\sim}) = \text{D-lim } \lambda i\ A_i(P_\eta)(f_1(i)), \ldots, f_{\tau(\eta)}(i)).$$

It is easily seen that each $\mathscr{A}(c_\zeta) \in R$. Moreover, since each $\tau(\eta)$ is finite, each $\mathscr{A}(P_\eta)$ is a well-defined function, independent of the representatives of the equivalence classes, mapping $R^{\tau(\eta)}$ into X. Thus A belongs to $\mathscr{M}$. Notice that the construction of D-prod F is independent of the $\mathscr{F}$ of the language $\mathscr{L}$ and depends only on X and the type of $\mathscr{L}$.

LEMMA 5.1.4. (The fundamental lemma). Let $A = \text{D-prod } F$, $T = \Pi_{i\in I} R_i$, $s \in T^\infty$, and $\varphi \in \Phi$. Define $r \in R^\infty$ by $r = (\lambda n{<}\omega)(s_n)^\sim$ and, for each $i \in I$, define $s_i = (\lambda n{<}\omega)s_n(i)$, so that $s_i \in (R_i)^\infty$. Then

$$(1) \qquad \varphi[\text{D-prod } F,\ r] = \text{D-lim } \lambda i\ \varphi[A_i,\ s_i].$$

PROOF. Let Ψ be the set of formulas for which (1) holds for all $s \in T^\infty$. One easily sees that $\Lambda \subset \Psi$. Assume that $\varphi_1, \ldots, \varphi_n \in \Psi$, $f \in \mathscr{C}_n \cap \mathscr{F}$, and $\varphi = f(\varphi_1, \ldots, \varphi_n)$. Then by 3.2.5, the inductive hypothesis, and Theorem 1.5.3,

$$\begin{aligned} \varphi[A,\ r] &= f(\varphi_1[A,\ r],\ \ldots,\ \varphi_n[A,\ r]) \\ &= f(\text{D-lim } \lambda i\ \varphi_1[A_i,\ s_i],\ \ldots,\ \text{D-lim } \lambda i\ \varphi_n[A_i,\ s_i]) \\ &= \text{D-lim } \lambda i\ f(\varphi_1[A_i,\ s_i],\ \ldots,\ \varphi_n[A_i,\ s_i]) \\ &= \text{D-lim } \lambda i\ \varphi[A_i,\ s_i]. \end{aligned}$$

So we have $\varphi \in \Psi$.

Assume now that $\psi \in \Psi$, $q \in \mathscr{F} \cap \mathscr{Q}$, and $\varphi = (qv_n)\psi$. By 3.2.6 and the inductive hypothesis,

$$\begin{aligned} \varphi[A,\ r\ ] &= q(\{\psi[A,\ r']\ :\ r' \in R^\infty \text{ and } r'_m = r_m \text{ for } m \neq n\}) \\ &= q(\{\text{D-lim } \lambda i\ \psi[A_i,\ s'_i]\ :\ s' \in T^\infty \text{ and } s'_m = s_m \text{ for } m \neq n\}). \end{aligned}$$

Let

$$Y = \{\lambda i\ \ \psi[A_i,\ s'_i]\ :\ s' \in T^\infty \text{ and } s'_m = s_m \text{ for } m \neq n\},$$

and let

$$Y_i = \{\psi[A_i,\ s'_i]\ :\ s' \in T^\infty \text{ and } s'_m = s_m \text{ for } m \neq n\}.$$

It follows that, for each $i \in I$,

$$Y_i = \{\psi[A_i, r'] : r' \in (R_i)^\infty \text{ and } r'_m = s_m(i) \text{ for } m \neq n\}.$$

Hence, $Y = \Pi_{i \in I} Y_i$, and by Theorem 1.5.6,

$$q(\text{D-lim } Y) = q(\text{D*-lim } \lambda i\ Y_i) = \text{D-lim } \lambda i\ q(Y_i).$$

Now, each

$$q(Y_i) = (qv_n)\psi[A_i, s_i] = \varphi[A_i, s_i],$$

so we have

$$\varphi[A, r] = \text{D-lim } \lambda i\ \varphi[A_i, s_i].$$

Thus $\varphi \in \Psi$. So $\Phi \subset \Psi$ and the lemma is proved.

We see that Lemma 5.1.4 continues to hold if the set of all formulas is enlarged to include $(\mathcal{C} \cup \mathcal{Q})\Lambda$. There are several places in this chapter where we prove a theorem for all $\varphi \in \Phi$, when we could equally well prove the theorem for all $\varphi \in (\mathcal{C} \cup \mathcal{Q})\Lambda$.

COROLLARY 5.1.5. For every sentence $\varphi \in \Sigma$,

$$\varphi[\text{D-prod } F] = \text{D-lim } \lambda i\ \varphi[A_i].$$

It follows trivially from Corollary 5.1.5. that if $[A_i] = [B_i]$ for each $i \in I$, then

$$[\text{D-prod } \lambda_i A_i] = [\text{D-prod } \lambda_i B_i].$$

Since X^Σ is a compact Hausdorff space, each function $h : I \rightarrow X^\Sigma$ has a D-limit in X^Σ. In terms of the space X^Σ we may state the following corollary.

COROLLARY 5.1.6. $[\text{D-prod } F] = \text{D-lim } \lambda i [A_i]$.

In case each $A_i = B$, that is, the function F is the constant function with value B, we simply write D-prod B for D-prod F, and we call D-prod B the *ultrapower of* B *modulo* D.

COROLLARY 5.1.7. [D-prod B] = [B], and thus B is elementarily equivalent to its ultrapower.

PROOF. By 5.1.6 and 1.5.2, as the D-limit of a constant function is the value of the function.

Turning to the extended theory of a model, we have the following important consequences of the fundamental lemma.

COROLLARY 5.1.8. For every function $f \in \Pi_{i \in I} R_i$ we have, in the space X^{Σ^+},

$$[\text{D-prod } F, f^{\sim}] = \text{D-lim } \lambda i \ [A_i, f(i)].$$

COROLLARY 5.1.9. $Th^+(\text{D-prod } F) = D^*\text{-lim } \lambda i \ Th^+(A_i)$.

PROOF. We observe that each of the following are equivalent:

$h \in Th^+(\text{D-prod } F)$;

for some $r_0 \in R$, $h = [\text{D-prod } F, r_0]$;

for some $f \in \Pi_{i \in I} R_i$, $h = \text{D-lim } \lambda i \ [A_i, f(i)]$;

for some $g \in \Pi_{i \in I} \ Th^+(A_i)$, $h = \text{D-lim } g$;

$h \in D^*\text{-lim } \lambda i \ Th^+(A_i)$.

THEOREM 5.1.10. Suppose that $\|\mathcal{L}\| \leq |J|$, $I = S_\omega(J)$, and D is a regular ultrafilter over I. Then, for every $B \in \mathcal{M}$,

$$Th^+(\text{D-prod } B) = \overline{Th^+(\text{D-prod } B)} = \overline{Th^+(B)}.$$

PROOF. The space X^{Σ^+} has an open base of power $\|\mathcal{L}\|$ and is compact and Hausdorff. Therefore we may apply Theorem 1.5.9. to the space X^{Σ^+} and we have,

$$\overline{Th^+(B)} = \text{D-lim } Th^+(B)^I = D^*\text{-lim } \lambda i \ Th^+(B) .$$

By Corollary 5.1.9,

$$D^*\text{-lim } \lambda i \ Th^+(B) = Th^+(\text{D-prod } B),$$

and the desired result follows.

EXERCISE 5A. Prove that for any D and B we have, in the space X^{Σ^+},

$$Th^+(\text{D-prod } B) \subset \overline{Th^+(B)}.$$

EXERCISE 5B. Obtain results analogous to Corollary 5.1.9 and Theorem 5.1.10 for Th^∞ and for Th^μ as defined in Exercise 4A.

EXERCISE 5C*. Suppose that $\mathcal{L}$ has a t-set, a k-set, and an e-set. Show that if $Th(A) = Th(B)$, then A is elementarily embeddable in some ultrapower of B. Hint: Use Exercises 4A and 5B.

EXERCISE 5D*. Show by means of a counterexample that Theorem 5.1.10 is no longer true if the hypothesis $\|\mathcal{L}\| \leq |J|$ is weakened to $\omega \cup |\mathcal{S}_0| \leq |J|$.

5.2. The compactness theorem

Let $K \subset \mathcal{M}$. We say that K is *closed under ultraproducts* if D-prod $F \in K$ for every non-empty set I, function $F \in K^I$, and ultrafilter D over I.

LEMMA 5.2.1. .1) If K is closed under ultraproducts, then $Th(K)$ is closed in the space X^Σ.
.2) If Θ is a closed theory, then Mod (Θ) is closed under ultraproducts.

PROOF. To prove .1), assume that K is closed under ultraproducts. By Theorem 1.5.10, it is sufficient to prove that $Th(K)$ is closed under D-limits. Let $G \in Th(K)^I$ and let D be an ultrafilter over I. Since $G \in Th(K)^I$, there exists an $F \in K^I$ such that $G(i) = [F(i)]$ for each $i \in I$. Now, D-prod $F \in K$ and, by 5.1.6, [D-prod F] = D-lim G, so D-lim $G \in Th(K)$.

To prove .2), let Θ be a closed set in X^Σ, and let $F \in$ Mod $(\Theta)^I$ and D be an ultrafilter over I. Since each $F(i) \in$ Mod (Θ), each $[F(i)] \in \Theta$. Since Θ is closed, we have D-lim λi $[F(i)] \in \Theta$. Using 5.1.6 once more, we have [D-prod F] $\in \Theta$, so D-prod $F \in$ Mod (Θ). The lemma is proved.

By combining 5.2.1.1) and Theorem 3.3.4, we see at once that any K which is closed under ultraproducts has the properties 3.3.4(i) and 3.3.4(ii).

The following theorem is the compactness theorem for continuous model theory.

THEOREM 5.2.2. The elementary topology is compact.

PROOF. Immediate from Lemma 5.2.1 and Exercise 3Q.6).

COROLLARY 5.2.3. Let $\varphi \in \Sigma$. Then the set $\{\varphi[A] : A \in \mathcal{M}\}$ is closed in X.

The following is a generalization of a well-known characterization of elementary classes in Frayne-Morel-Scott [1962].

THEOREM 5.2.4. $K \in EC_{\Delta}$ if and only if K is closed under ultraproducts and K is elementarily closed.

PROOF. Assume $K \in EC_{\Delta}$. By Exercise 3Q.4), K is eletarily closed. By Exercise 3Q.1), $K = \text{Mod}\ \overline{\text{Th}(K)}$. Hence by Lemma 5.2.1.2), K is closed under ultraproducts.

On the other hand, if K is elementarily closed and closed under ultraproducts, then, by Lemma 5.2.1.1), $\text{Th}(K)$ is closed in X^{Σ}, and, by Exercise 3Q.5), $K \in EC_{\Delta}$. The theorem is proved.

EXERCISE 5E. Let $K \subset \mathcal{M}$. Then the class

$$\{A \in \mathcal{M} : \text{Th}(A) = \text{Th}(D\text{-prod } F) \text{ for some } F \in K^I \text{ and ultrafilter } D \text{ over } I\}$$

is an elementary class. Hence, if K is closed under ultraproducts, then the class

$$\{A \in \mathcal{M} : \text{Th}\ (A) = \text{Th}\ (B) \text{ for some } B \in K\}$$

is an elementary class.

EXERCISE 5F. Let $K \subset \mathcal{M}$. Then both K and $\mathcal{M} - K$ are finite intersections of basic elementary classes if and only if

(*) both K and $\mathcal{M} - K$ are closed under ultraproducts and K is elementarily closed.

Furthermore, if $\mathcal{L}$ has a t-set and a k-set, then both K and $\mathcal{M} - K$ are finite intersections of singular elementary classes if and only if (*) holds. In the case of the logic $\mathfrak{L}$, both K and $\mathcal{M} - K$ are singular elementary classes if and only if (*) holds.

5.3. The upward Löwenheim-Skolem theorem

In this section we continue to assume that I is a nonempty set and D is an ultrafilter over I. By the *natural embedding of* $\mathfrak{B}$ *into* D-prod $\mathfrak{B}$ we shall mean the function which maps each element $b \in S$ into the element $(\lambda i\ b)^{\sim} \in D\text{-prod } S$.

THEOREM 5.3.1. The natural embedding is an isomorphism of B onto an elementary submodel of D-prod B.

PROOF. Let B' be the submodel of D-prod B generated by the range of the natural embedding, and denote the natural embedding by d. The proof that d is an isomorphism of B onto B' is trivial.

Let $\|B\| = \beta$ and let $b \in S^{\beta}$ be an enumeration of S. Applying Corollary 5.1.7 in the language $\mathcal{L}(\beta)$, we have

$$\mathrm{Th}((B, b)) = \mathrm{Th}(\text{D-prod}(B, b)).$$

From the definition of D-prod(B, b), we see that

$$\text{D-prod}(B, b) = (\text{D-prod } B, d \circ b).$$

The range of $d \circ b$ is the set S' of elements of B'. Now since $(B', d \circ b) \cong (B, b)$, we have

$$\mathrm{Th}((B', d \circ b)) = \mathrm{Th}((\text{D-prod } B, d \circ b)).$$

By Theorem 4.2.7, this means that B' is an elementary submodel of D-prod B. The theorem is proved.

EXERCISE 5G. If either B is finite or I is finite, then D-prod B $\cong$ B.

EXERCISE 5H. Show that if B is infinite and D is countably incomplete, then the natural embedding maps B onto a proper submodel of D-prod B.

EXERCISE 5I*.1) Show that each model B is isomorphically embeddable in some ultraproduct D-prod $\lambda i\, A_i$ where each A_i is a finite model. Furthermore, if κ is finite, then each A_i can be taken to be a finite submodel of B. Can B be elementarily embedded in such an ultraproduct?

.2) Let Θ be a consistent theory and let $\|\mathcal{L}\| \leq \alpha$. Suppose that any two models of Θ can be isomorphically embedded in a third model of Θ. Then show that for any collection A_η, $\eta < \alpha$, of models of Θ, each with power α, there exists a model A of Θ with power α, such that each A_η, $\eta < \alpha$, can be isomorphically embedded in A.

EXERCISE 5J*.1) Suppose $B \prec A$. Then there are an elementary embedding

$$h : \text{D-prod } B \to \text{D-prod } A$$

and two natural embeddings

$$d_1 : B \to \text{D-prod } B,$$

$$d_2 : A \to \text{D-prod } A,$$

such that $h \circ d_1 = d_2 \restriction S$. Hence, there are models A' and B' such that

$$A' \cong \text{D-prod } A, \quad B' \cong \text{D-prod } B,$$

$$B \prec B' \prec A', \quad B \prec A \prec A'.$$

.2) Suppose that $B = \bigcup_{\eta<\alpha} A_\eta$, where A_η, $\eta < \alpha$, is a chain of models. Then for some ultrafilter D over α, there are isomorphic embeddings

$$h_1 : B \to \text{D-prod } \lambda\eta\, A_\eta, \quad h_2 : \text{D-prod } \lambda\eta\, A_\eta \to \text{D-prod } B$$

such that $h_2 \circ h_1$ is the natural embedding of B into D-prod B. Infer from this that if

(i) K is closed under ultraproducts and isomorphisms, and

(ii) $A \in K$, $B \subset A \subset C$, and $B \prec C$ imply $B \in K$, then

(iii) K is closed under unions of chains of models in K.

We now prove the "upward Löwenheim-Skolem theorem" for continuous logics.

THEOREM 5.3.2. Let A be an infinite model and let β be a cardinal such that $\|\mathcal{L}\| \leq \beta$ and $\|A\| \leq \beta$. Then A has an elementary extension of power β.

PROOF. By Lemma 1.2.2. A has an ultrapower of power $\geq \beta$. By Theorem 5.3.1 every ultrapower of A is isomorphic to an elementary extension of A, and so A has an elementary extension of power $\geq \beta$. By 4.3.1 we can find an elementary extension of A of power β.

COROLLARY 5.3.3. Suppose the theory Θ has an infinite model. Then, for every cardinal $\beta \geq \|\mathcal{L}\|$, Θ has a model of power β.

PROOF. This follows from 4.2.6 and 5.3.2.

EXERCISE 5K. Suppose that $\beta \geq \| \mathcal{L} \|$. and that Θ is a theory which has only infinite models. If any two models of Θ of power β are elementarily equivalent, then Θ is complete. Furthermore, if any two models A and B of Θ of power β satisfy

$$(*) \qquad \overline{\mathrm{Th}^+(A)} = \overline{\mathrm{Th}^+(B)},$$

then any two models A and B of Θ satisfy (*).

EXERCISE 5L*. In contrast to Exercise 4G, prove that Corollary 5.3.3 is in general not true if X is not assumed to be compact. Suppose that $|\Phi| = \omega$ and X is a countable discrete space. There is a least cardinal β such that whenever Θ has a model of power β then it has models of power γ for every $\gamma \geq \beta$. (How large is β?)

5.4 Good ultrafilters

In this section we introduce the notion of an α-good ultrafilter and prove some purely set-theoretical results concerning them. Good ultrafilters could have been introduced in Chapter I, but it appears better to us to introduce them at this point, so that we can establish their model-theoretic significance at once in the next section.

Let I be a non-empty set and let α be an infinite cardinal. If f and g are any two functions, we shall write $f \leq g$ if $\mathcal{D}f = \mathcal{D}g$ and $f(u) \subset g(u)$ for all $u \in \mathcal{D}f$. Consider a function f on $S^{\omega}(\alpha)$ into $S(I)$. We shall say that f is *monotonic* if

$$f(u) \subset f(w) \quad \text{whenever} \quad u, w \in \mathcal{D}f \quad \text{and} \quad u \subset w.$$

We say that f is *multiplicative* if

$$f(u \cap w) = f(u) \cap f(w) \quad \text{for all} \quad u, w \in \mathcal{D}f.$$

Obviously, any multiplicative function is monotonic.

An ultrafilter D over I is said to be *α-good* if it has the following property:

For every $\beta < \alpha$ and every monotonic function

$f: S^{\omega}(\beta) \to D$, there exists a multiplicative function $g: S^{\omega}(\beta) \to D$ such that $g \leq f$.

EXERCISE 5M. Every principal ultrafilter D is α-good for all α.

Notice that if D is α-good, then D is β-good for all $\beta < \alpha$.

We recall from Exercise 1 J that D is said to be *countably incomplete* if there exists a countable subset $E \subset D$ such that $\bigcap E \notin D$. It is known that unless the I is of exceedingly large cardinality, every ultrafilter D over I is either principal or countably incomplete.

LEMMA 5.4.1. If D is countably incomplete, then there is a countable decreasing sequence $y_0 \supset y_1 \supset y_2 \supset \ldots$ of elements of D such that $I = y_0$ and $\bigcap_{n<\omega} y_n = 0$.

PROOF. Let E be a countable subset of D with $\bigcap E \notin D$. We may write .

$$E = \{e_n : n < \omega\}.$$

Let $y_0 = I$, and for each n, define

$$y_{n+1} = y_n \cap e_n \cap (I - \bigcap E).$$

The sequence $y_0, y_1, y_2, \ldots$ clearly has the desired properties.

THEOREM 5.4.2. Every ultrafilter D is ω^+-good.

PROOF. Let $\beta < \omega^+$, and let $f: S^{\omega}(\beta) \to D$ be monotonic. If β is finite, then $0 \in S^{\omega}(\beta)$ and the constant function

$$g = (\lambda u \in S^{\omega}(\beta)) f(0)$$

is multiplicative, has values in D, and is such that $g \leq f$.

The remaining case is $\beta = \omega$. For each $u \in S^{\omega}(\omega)$, let $m(u)$ be the smallest natural number m such that $\omega - m \subset u$. We define

$$g = \lambda u\ f(\omega - m(u)).$$

Then $g\colon S^{\omega}(\omega) \to D$, $g \leq f$, and since $m(u \cap w) = m(u) \cup m(w)$, g is multiplicative. Our proof is complete.

To prove the existence of α-good ultrafilters for $\alpha > \omega^+$, we must go to considerable lengths. This we proceed to do.

THEOREM 5.4.3. Assume the generalized continuum hypothesis. Let $|I| = \alpha$. Then there exists an α^+-good, countably incomplete, ultrafilter D over I.

To facilitate the proof, we shall first prove a series of lemmas which do not depend on the generalized continuum hypothesis.

LEMMA 5.4.4. For D to be α^+-good it is necessary and sufficient that for every monotonic function $f\colon S^{\omega}(\alpha) \to D$ there is a multiplicative function $g\colon S^{\omega}(\alpha) \to D$ such that $g \leq f$.

PROOF. The necessity is obvious. To prove the sufficiency, let $\beta \leq \alpha$, and let $f\colon S^{\omega}(\beta) \to D$ be monotonic. Define

$$f' = (\lambda u \in S^{\omega}(\alpha))f(u \cap \beta).$$

Then f' is monotonic. By hypothesis there exists a multiplicatine function $g'\colon S^{\omega}(\alpha) \to D$ with $g' \leq f'$. Now define

$$g = (\lambda u \in S^{\omega}(\beta))g'(u \cup (\alpha - \beta)).$$

Then g is multiplicative, $g \leq f$, and the values of g are in D. Hence D is α^+-good.

A function k is said to be *disjointed* if

$$k(i) \cap k(j) = 0 \quad \text{whenever} \quad i, j \in \mathscr{D}k \quad \text{and} \quad i \neq j.$$

LEMMA 5.4.5. Let h be a function such that $|\mathscr{D}\ h| = \alpha$ and $|h(j)| = \alpha$ for all $j \in \mathscr{D}h$. Then there exists a disjointed function k such that $k \leq h$ and $|k(j)| = \alpha$

for all $j \in \mathcal{D}h$.

PROOF. We may assume without loss of generality that $\mathcal{D}h = \alpha$. For each $\zeta \leq \alpha$, let

$$x_\zeta = \{\langle\eta, \xi\rangle : \eta \leq \xi \text{ and } \xi < \zeta\}.$$

We wish to find a one-one function e such that $\mathcal{D}e = x_\alpha$ and $e(\eta,\xi) \in h(\eta)$ whenever $\eta \leq \xi < \alpha$. Such a function may be found by transfinite induction, as follows. Let $\zeta < \alpha$ and suppose that we already have found a one-one function e_0 with domain x_ζ such that $e_0(\eta, \xi) \in h(\eta)$ whenever $\eta \leq \xi < \zeta$. We have $|x_\zeta| < \alpha$. Therefore, since $|h(\eta)| = \alpha$ for all $\eta < \alpha$, we may extend e_0 to a one-one function e_1 with domain $x_{\zeta+1}$ such that $e_1(\eta, \xi) \in h(\eta)$ whenever $\eta \leq \xi \leq \zeta$. The existence of the desired function e follows.

We now define k by:

$$k = (\lambda\eta < \alpha)\{e(\eta,\xi) : \langle\eta,\xi\rangle \in x_\alpha\}.$$

Obviously, k is disjointed and $k \leq h$. Moreover, for each $\eta < \alpha$, we have

$$|k(\eta)| = |\alpha - \eta| = \alpha.$$

A set $E \subset S(I)$ is said to be *multiplicative* if the intersection of any two members of E belongs to E.

LEMMA 5.4.6. Suppose that E is a multiplicative subset of $S(I)$ such that $|E| \leq \alpha$, and for each $e \in E$, $|e| = \alpha$. Let $f\colon S^\omega(\alpha) \to S(I)$ be a monotonic function such that

$$|f(u) \cap e| = \alpha \quad \text{for all } u \in \mathcal{D}f \text{ and } e \in E.$$

Then there is a multiplicative function $g\colon S^\omega(\alpha) \to S(I)$ such that $g \leq f$ and

$$|g(u) \cap e| = \alpha \quad \text{for all } u, e.$$

PROOF. Let h be the function on $S^{\omega}(\alpha) \times E$ into $S(I)$ such that, for each $u \in S^{\omega}(\alpha)$ and $e \in E$,

$$h(u, e) = f(u) \cap e.$$

Then by hypothesis we always have

$$|h(u, e)| = \alpha.$$

Since α is infinite the domain of h also has power α. Thus by Lemma 5.4.5 there exists a disjointed function $k \leq h$ such that, whenever $\langle u, e\rangle \in \mathcal{D}k$, $|k(u, e)| = \alpha$. We shall define the function $g: S^{\omega}(\alpha) \to S(I)$ as follows: for each $u \in S^{\omega}(\alpha)$,

$$(1) \qquad g(u) = \bigcup\{k(w, e) : w \in S^{\omega}(\alpha),\ w \subset u,\ \text{and}\ e \in E\}.$$

We shall show that g has the desired properties.

Suppose $u \in S^{\omega}(\alpha)$ and $i \in g(u)$. Then for some $w \subset u$ and $e \in E$, we have

$$i \in k(w, e) \subset h(w, e) = f(w) \cap e.$$

Since f is monotonic, $f(w) \subset f(u)$, and so $i \in f(u)$. This verifies that $g \leq f$.

To show that g is multiplicative, we observe that for each $u_1, u_2 \in S^{\omega}(\alpha)$ and $i \in I$, the following are equivalent:

(a) $i \in g(u_1 \cap u_2)$;

(b) $i \in k(w, e)$ for some $w \subset u_1 \cap u_2$ and $e \in E$;

(c) $i \in k(w_1, e_1) \cap k(w_2, e_2)$ for some $w_1 \subset u_1$, $w_2 \subset u_2$ and $e_1, e_2 \in E$;

(d) $i \in g(u_1) \cap g(u_2)$.

That (c) implies (b) follows from the fact that, if $i \in k(w_1, e_1) \cap k(w_2, e_2)$, then $w_1 = w_2$ and $e_1 = e_2$.

Let $u \in S^{\omega}(\alpha)$ and $e \in E$. We show that $|g(u) \cap e| = \alpha$. By (1) we have $k(u, e) \subset g(u)$. Since $k \leq h$,

$$k(u, e) \subset h(u, e) = f(u) \cap e.$$

Therefore

$$k(u, e) \subset g(u) \cap e.$$

Since $|k(u, e)| = \alpha$, we have $|g(u) \cap e| = \alpha$.

PROOF OF THEOREM 5.4.3. We first take steps to insure that the ultrafilter D which we shall construct is countably incomplete. Accordingly, let $y_0 \supset y_1 \supset y_2 \supset \ldots$ be a countable decreasing sequence of sets such that $y_0 = I$, $\bigcap_{n<\omega} y_n = 0$, and each y_n is a power of α. Let

$$D_0 = \{y_n : n < \omega\}.$$

D_0 is clearly multiplicative and of power $\omega \leq \alpha$.

Since $|S^{\omega}(\alpha)| = \alpha$, there are at most $(2^{\alpha})^{\alpha} = 2^{\alpha}$ monotonic functions f on $S^{\omega}(\alpha)$ into $S(I)$. By hypothesis, $2^{\alpha} = \alpha^{+}$, so there are at most α^{+} such functions. Let $(\lambda\eta < \alpha^{+})f_{\eta}$ be an enumeration of all monotonic functions $f\colon S^{\omega}(\alpha) \to S(I)$. By Lemma 5.4.6, we may choose, for each $\eta < \alpha^{+}$ and each multiplicative set $E \subset S(I)$ of power $\leq \alpha$ satisfying the condition

(1) $|f_{\eta}(u) \cap e| = \alpha$ for all $u \in S^{\omega}(\alpha)$ and $e \in E$, a multiplicative function $g_{\eta,E}$ such that $g_{\eta,E} \leq f_{\eta}$ and

(2) $|g_{\eta,E}(u) \cap e| = \alpha$ for all $u \in S^{\omega}(\alpha)$ and $e \in E$. We define a sequence $(\lambda\eta < \alpha^{+})E_{\eta}$ by transfinite recursion as follows:

(3) $E_0 = D_0$;

(4) If ζ is a positive limit ordinal $< \alpha^{+}$, then $E_{\zeta} = \bigcup_{\eta<\zeta} E_{\eta}$;

(5) If $\eta < \alpha^{+}$ and (1) fails with $E = E_{\eta}$, then $E_{\eta+1} = E_{\eta}$;

(6) If $\eta < \alpha^{+}$ and (1) holds with $E = E_{\eta}$, then $E_{\eta+1} = E_{\eta} \cup \{g_{\eta,E_{\eta}}(u) \cap e : u \in S^{\omega}(\alpha),\ e \in E_{\eta}\}$.

It is easy to show by induction that each set E_η is of power $\leq \alpha$, that each $e \in E_\eta$ is of power α, and that the sequence $(\lambda\eta < \alpha^+)E_\eta$ exists. Let

$$E' = \bigcup_{\eta<\alpha^+} E_\eta .$$

Then each $e \in E'$ is of power α, and E' is multiplicative. Hence there exists an ultrafilter D over I such that $E' \cup S^\alpha(I) \subset D$. D is countably incomplete because $D_0 \subset E' \subset D$.

Let f be any monotonic function on $S^\omega(\alpha)$ into D. Then $f = f_\eta$ for some $\eta < \alpha^+$. Since $S^\alpha(I) \subset D$, each member of D has power α, and thus since $E_\eta \subset D$, (1) holds with $E = E_\eta$. Then by (2) and (6), the multiplicative function $g_{\eta,E_\eta} \leq f_\eta$ maps $S^\omega(\alpha)$ into $E_{\eta+1}$. Hence g_{η,E_η} maps $S^\omega(\alpha)$ into D. Therefore, by Lemma 5.4.4, D is α^+-good, and our proof is complete.

EXERCISE 5N*. Suppose $|I| = \alpha$ and D is an α^+-good, countably incomplete, ultrafilter over I. Then

(i) $S^\alpha(I) \subset D$;

(ii) D is weakly regular.

EXERCISE 5O. Suppose $|I| = \alpha$ and $2^\alpha = \alpha^+$. By examining the proof of Theorem 5.4.3, show that any set E as in Lemma 5.4.6 can be extended to an α^+-good, countably incomplete ultrafilter D over I. Conclude from this that, if $I = S_\omega(J)$, then:

(i) there exists a regular α^+-good ultrafilter D over I; and

(ii) there exists an α^+-good ultrafilter D' over I which is not regular.

5.5. Good ultraproducts

In this section we shall apply good ultrafilters to continuous model theory in a way which is analogous to the applications of regular ultrafilters which we have already given. Our first theorem is a purely topological result which is the analog of Theorem 1.5.9.

THEOREM 5.5.1. Suppose that $|\mathcal{S}_0| < \alpha$ and D is an α-good, countably incomplete ultrafilter over I. Then for any function $F \in (X^*)^I$, the set $D^*\text{-lim } F$ is closed in X.

PROOF. Let $x \in \overline{D^*\text{-lim } F}$. It suffices to show that there is a function $h \in \Pi_{i\in I} F(i)$ such that $x = D\text{-lim } h$. Let

$$T = \{U \in \mathcal{S}_0 : x \in U\} .$$

Let us define the function $f\colon S^{\omega}(T) \rightarrowtail S(I)$ as follows:

$$f(T - Z) = \{i \in I : F(i) \cap \bigcap Z \neq 0\} ,$$

where Z is any finite subset of T. It is obvious that f is monotonic. for each $Z \in S_{\omega}(T)$, $\bigcap Z$ is a neighborhood of the point x, and therefore $\bigcap Z$ meets the set $D^*\text{-lim } F$. This means that there exists $h_Z \in \Pi_{i\in I} F(i)$ such that $D\text{-lim } h_Z \in \bigcap Z$, and hence

$$\{i \in I : h_Z(i) \in \bigcap Z\} \in D .$$

It follows that $f(T - Z) \in D$ for each $T - Z \in S^{\omega}(T)$, that is, $f\colon S^{\omega}(T) \rightarrow D$.

Since D is countably incomplete, there is by Lemma 5.4.1 a countable decreasing sequence $y_0 \supset y_1 \supset y_2 \supset \cdots$ of elements of D such that $y_0 = I$ and $\bigcap_{n<\omega} y_n = 0$. Let f' be the function defined by

$$f'(T - Z) = f(T - Z) \cap y_{|Z|},$$

for each $Z \in S_{\omega}(T)$. It is clear that $f' \leq f$, f' is monotonic, and f' maps $S^{\omega}(T)$ into D.

Now, using the hypothesis that D is α-good and $|T| \leq |\mathcal{S}_0| < \alpha$, we may choose a multiplicative function $g\colon S^{\omega}(T) \rightarrow D$ such that $g \leq f'$. It is clear that g may be chosen so that $g(T) = I$, because our assumption $x \in \overline{D^*\text{-lim } F}$ implies that each $F(i) \neq 0$ and hence that $f(T) = f'(T) = I$. For each $i \in I$, let $n(i)$ be the greatest $n < \omega$ such that $i \in y_n$, and let

$$Z(i) = \{U \in T : i \in g(T - \{U\})\}.$$

The power of $Z(i)$ is at most $n(i)$; for otherwise there would be a subset $Z_0 \subset Z(i)$ of power $n(i) + 1$, and by the multiplicativeness of g we would have

$$i \in \bigcap_{U \in Z_0} g(T - \{U\}) = g(T - Z_0) \subset f'(T - Z_0) \subset y_{n(i)+1} ,$$

contradicting $i \notin y_{n(i)+1}$. Hence for all $i \in I$,

$$T - Z(i) \in S^{\omega}(T) \quad \text{and} \quad i \in g(T - Z(i)).$$

Since $g \leq f$, we have $i \in f(T - Z(i))$, and therefore

$$F(i) \cap \bigcap Z(i) \neq 0 \quad \text{for each} \quad i \in I,$$

putting $\bigcap 0 = X$. Choose $h \in X^I$ such that

$$h(i) \in F(i) \cap \bigcap Z(i) \quad \text{for all} \quad i \in I.$$

Then $h \in \Pi_{i \in I} F(i)$. Moreover, for each $x \in U \in \mathscr{S}_0$,

$$g(T - \{U\}) = \{i \in I : U \in Z(i)\} \subset \{i \in I : h(i) \in U\}$$

and hence

$$\{i \in I : h(I) \in U\} \in D.$$

It follows that $x = D\text{-lim } h$, and our proof is complete.

We now prove the analog of Theorem 5.1.10.

THEOREM 5.5.2. Suppose that $\|\mathscr{L}\| < \alpha$ and D is an α-good, countably incomplete ultrafilter over I. Then, for every function $\lambda i\ A_i \in \mathscr{M}^I$, the extended theory $Th^+(D\text{-prod } \lambda i\ A_i)$ of the ultraproduct is closed in X^{Σ^+}.

PROOF. The space X^{Σ^+} has an open base of power $\|\mathscr{L}\| < \alpha$ and is compact and Hausdorff. Therefore we may apply Theorem 5.5.1 to the space X^{Σ^+} and we see that the set $D^*\text{-lim } \lambda i\ Th^+(A_i)$ is closed in X^{Σ^+}. By Corollary 5.1.9,

$$D^*\text{-lim } \lambda i\ Th^+(A_i) = Th^+(D\text{-prod } \lambda i\ A_i) ,$$

and the result follows.

The importance of Theorem 5.5.2 above, and also of Theorem 5.1.10, becomes evident when we introduce the notion of an α-saturated model, which will be the subject matter of the next chapter. Anticipating the next chapter, we shall give the definition at this time.

We say that a model $A \in \mathcal{M}$ is *α-saturated* if, for every $\upsilon < \alpha$ and every sequence $a \in R^{\upsilon}$, the set $Th^{+}((A, a))$ is closed in $X^{\Sigma^{+}_{\mathcal{L}(\upsilon)}}$.

LEMMA 5.5.3. Let $A = \text{D-prod } \lambda i\ A_i$ be an ultraproduct, let υ be an ordinal, and let $a \in R^{\upsilon}$. Then there exists a function $b \in \Pi_{i\in I}(R_i^{\upsilon})$ such that

$$(A, a) = \text{D-prod } \lambda i\ (A_i, b_i).$$

PROOF. For each $\eta < \upsilon$, choose $f_\eta \in a_\eta$, that is, $f_\eta \in \Pi_{i\in I} R_i$ and $f_\eta^{\sim} = a_\eta$. Define b so that, for each $i \in I$,

$$b_i = (\lambda\eta < \upsilon) f_\eta(i).$$

Then for each $\eta < \upsilon$ we have

$$a_\eta = f_\eta^{\sim} = (\lambda i\ b_{i\eta})^{\sim},$$

and thus b has the required property.

THEOREM 5.5.4. Suppose that $\|\mathcal{L}\| < \alpha$ and D is an α-good, countably incomplete ultrafilter over I. Then every D-product $A = \text{D-prod } \lambda i\ A_i$ is α-saturated.

PROOF. Let $\upsilon < \alpha$. Then $\|\mathcal{L}(\upsilon)\| < \alpha$. Let $A = \text{D-prod } \lambda i\ A_i$ and let $a \in R^{\upsilon}$. By Lemma 5.5.3, there exist $b_i \in R_i^{\upsilon}$ for each $i \in I$ such that

$$(A, a) = \text{D-prod } \lambda i\ (A_i, b_i).$$

Then by Theorem 5 5 2 applied to the logic $\mathcal{L}(\upsilon)$, the set $Th^+((A, a))$ is closed in $X^{\Sigma^+_{\mathcal{L}(\upsilon)}}$. It follows that A is α-saturated.

COROLLARY 5.5.5. If $\|\mathcal{L}\| = \omega$ and D is a countably incomplete ultrafilter over I, then every D-product $A = \text{D-prod } \lambda i\, A_i$ is ω^+-saturated.

PROOF. By Theorems 5.4.2 and 5.5.4:

COROLLARY 5.5.6. Assume the generalized continuum hypothesis. Suppose $\|\mathcal{L}\| \leq \alpha$ and $\|A\| \leq \alpha^+$. Then there is an α^+-saturated model B of power $\leq \alpha^+$ such that $A \prec B$.

PROOF. By Theorem 5.4.3., there exists an α^+-good, countably incomplete ultrafilter D over α. By 5.5.4, the model D-prod A is α^+-saturated. By Exercise 1F, $\|\text{D-prod } A\| \leq \|A\|^\alpha = \alpha^+$. Finally, by Theorem 5.3.1, there exists $B \cong \text{D-prod } A$ such that $A \prec B$.

We shall show in the next chapter that Corollary 5.5.6 above has a version which is valid even without the generalized continuum hypothesis. Instead of a single use of Theorem 5.5.4 and an α^+-good ultrafilter in the proof, we shall make repeated uses of Theorem 5.1.10 and a regular ultrafilter, and in this way the generalized continuum hypothesis will be avoided. On the other hand, in the last section of Chapter VI we shall obtain some results in which α^+-good ultrafilters and the generalized continuum hypothesis appear to be needed in an essential way.

EXERCISE 5P. Find the mistake in the following incorrect argument.

.1) Suppose $\|\mathcal{L}\| \leq |J| = \alpha$, $I = S_\omega(J)$, and D is a regular ultrafilter over I. Then for every model B, the ultrapower $A = \text{D-prod } B$ is α-saturated.

PROOF. Let $\upsilon < \alpha$ and $a \in R^{\upsilon}$. By 5.5.3 there exists $b \in S^{\upsilon}$ such that $(A, a) =$ D-prod (B, b). Hence by 5.1.10, and since $\| \mathcal{L}(\upsilon) \| < \alpha$,

$$\mathrm{Th}^{+}((A, a)) = \overline{\mathrm{Th}^{+}((A, a))} = \overline{\mathrm{Th}^{+}((B, b))}.$$

Thus A is α-saturated.

Indeed, it is known (cf. Keisler [1964a]) that the statement .1) of the above exercise is actually false, and counterexamples are known for each $\alpha > \omega^{+}$ in the classical two-valued logic ℓ.

CHAPTER VI

SPECIAL MODELS

We have already seen in §5.5 how the results on ultraproducts in §5.1-5.3 coupled with the results on good ultrafilters in §5.4 yield the existence of α^+-saturated models of power α^+ (Corollary 5.5.6, assuming the generalized continuum hypothesis). The object of this chapter is to study models which are, loosely speaking, unions of elementary chains where each term of the chain possesses a sufficiently high degree of saturatedness. We can prove, for instance, the existence of α^+-saturated models of power 2^α (without assuming the generalized continuum hypothesis). Using this result, we shall prove the existence of special models (§6.2) in certain infinite powers. Because of its many applications, the notion of a special model has already become very important in the two valued logic ℓ and its model theory. As we shall see, it plays an equally prominent role in continuous model theory. In fact, while many results in Chapter VII can be proved in two-valued model theory without using special models, their only known proofs in continuous model theory use special models.

The results of §6.1-6.2 hold for arbitrary continuous logics, while in §6.3-6.5 we shall need the assumption that $\mathcal{L}$ has a t-set, k-set, and e-set.

6.1. Saturated models

We recall from §5.5 that a model A is *α-saturated* (with respect to $\mathcal{L}$) if for every $\nu < \alpha$ and every $a \in R^\nu$, the set $Th^+((A, a))$ is closed in $X^{\Sigma_{\mathcal{L}(\nu)}^+}$. Whenever $\mathcal{L}$ and ν are understood, we shall simply say that A is α-saturated and that $Th^+((A, a))$ is closed.

We next state a number of simple properties of saturated models whose proofs we leave to the reader.

6.1.1. A is α-saturated if and only if A is β-saturated for all $\beta \leq \alpha$.

6.1.2. If α is a limit cardinal, then A is α-saturated if and only if A is β-saturated for all $\beta < \alpha$.

6.1.3. A is α^+-saturated if and only if $Th^+((A, a))$ is closed for all $a \in R^\alpha$.

6.1.4. The following are equivalent for $\alpha \geq \omega$:

(i) A is α-saturated.

(ii) For all $\nu < \alpha$ and $a \in R^\nu$, (A, a) is α-saturated with respect to $\mathcal{L}(\nu)$.

(iii) For all $\nu < \alpha$ and $a \in R^\nu$, (A, a) is 1-saturated with respect to $\mathcal{L}(\nu)$.

6.1.5. If A is α-saturated with respect to $\mathcal{L}$ and if $\mathcal{F}' \subset \mathcal{F}$, then A is also α-saturated with respect to $\mathcal{L}(\mathcal{F}')$.

EXERCISE 6A. Show that in the logic ℓ, a model $A \in \mathcal{M}_\ell$ is α-saturated if and only if for every $\nu < \alpha$, every $a \in R^\nu$, and every set of formulas $\Psi \subset {\Sigma_{\ell(\nu)}}^+$, Ψ is simultaneously satisfiable in (A, a) (by some element of A) whenever every finite subset of Ψ is satisfiable in (A, a).

EXERCISE 6B. Assume that $\mathcal{L}$ has a t-set, k-set, and e-set.

(i) If $A \prec B$ and A is 1-saturated, then B is 1-saturated. Hence the union of an elementary chain of 1-saturated models is 1-saturated.

(ii) The union of an elementary chain of ω-saturated models is ω-saturated.

(iii) The union of an elementary chain A_η, $\eta < \alpha^+$, of α^+-saturated models is α^+-saturated.

EXERCISE 6C*. Provide the following examples in the logic ℓ.

(i) A is 1-saturated but not 2-saturated.

(ii) $A \prec B$, A is ω-saturated, and B is not 2-saturated.

(iii) A union of a countable elementary chain of ω^+-saturated models which is not ω^+-saturated.

EXERCISE 6D*. Suppose that A is β^+-saturated. Then for every α such that $\|\mathcal{L}\| \leq \beta \leq \alpha^\beta \leq \|A\|$ and every $T \subset R$ such that $|T| \leq \alpha^\beta$, there exists a β^+-saturated elementary submodel of A of power α^β and containing T as a subset. (Hint: Since $|\mathcal{S}_0| \leq \beta$, the closure of a set $Y \subset X$ of power α^β is again of power α^β.)

LEMMA 6.1.6. If A is α-saturated and infinite, then $\|A\| \geq \alpha$.

PROOF. Suppose $\|A\| = \beta < \alpha$. Let $a \in R^\beta$ be an enumeration of R, and let

$$\Psi = \{v_0 \equiv c_{\kappa+\xi} : \xi < \beta\} \quad .$$

Then since R is infinite, the constant function $\underset{\sim}{0}$ belongs to

$$\overline{Th^+((A, a)) \upharpoonright \Psi}$$

but not to $Th^+((A, a)) \upharpoonright \Psi$. Hence the latter is not closed and neither is $Th^+((A, a))$.

THEOREM 6.1.7. A is finite if and only if A is α-saturated for all α.

PROOF. By Lemma 6.1.6, if A is infinite, then A is not $\|A\|^+$-saturated. On the other hand, suppose A is finite. Let β be a cardinal and let $a \in R^\beta$. $Th^+((A, a))$ is closed because it is finite.

LEMMA 6.1.8. Suppose $\|\mathcal{L}\| \leq \alpha$ and $\omega \leq \|A\| \leq 2^\alpha$. Then there exists an elementary extension B of A of power 2^α such that for all $a \in R^\alpha$,

$$Th^+((B, a)) = \overline{Th^+((B, a))} = \overline{Th^+((A, a))}.$$

PROOF. Let D be a regular ultrafilter on α, let $B' =$ D-prod A, and let A' be the image of A under the natural embedding d of A into B'. By Theorem 5.3.1, $A' \prec B'$, and by Exercise 1H, $\|B'\| = 2^\alpha$. Let $a' \in (R')^\alpha$. Then, for some $a \in R^\alpha$, $a' = d \circ a$. By Theorem 5.1.10 applied to the language $\mathcal{L}(\alpha)$ (notice that $\|\mathcal{L}(\alpha)\| \leq \alpha$), we have

$$\mathrm{Th}^+(\text{D-prod }(A, a)) = \overline{\mathrm{Th}^+(\text{D-prod }(A, a))} = \overline{\mathrm{Th}^+((A, a))}.$$

Since D-prod $(A, a) = (\text{D-prod } A, d \circ a) = (B', a')$ and since $(A, a) \cong (A', a')$, we have

$$\mathrm{Th}^+((B', a')) = \overline{\mathrm{Th}^+((B', a'))} = \overline{\mathrm{Th}^+((A', a'))} .$$

By a well-known set-theoretical construction, we can find a model B satisfying the conclusions of the lemma.

The following is a version of Corollary 5.5.6.

THEOREM 6.1.9. Suppose $\|\mathcal{L}\| \leq \alpha$ and $\omega \leq \|A\| \leq 2^\alpha$. Then A has an α^+-saturated elementary extension of power 2^α.

PROOF. We define a transfinite sequence of models A_η, $\eta \leq \alpha^+$, as follows:

$A_0 = A$;

$A_{\eta+1}$ is a model of power 2^α such that $A_\eta \prec A_{\eta+1}$ and for every $a \in R_\eta^\alpha$,

$$\mathrm{Th}^+((A_{\eta+1}, a)) = \overline{\mathrm{Th}^+((A_{\eta+1}, a))} = \overline{\mathrm{Th}^+((A_\eta, a))} ;$$

$A_\eta = \bigcup_{\xi < \eta} A_\xi$ if η is a limit ordinal different from 0.

Clearly, by Lemma 6.1.8, such a sequence of models exists and it is an elementary chain. By an easy induction, we have

$$\|A_\eta\| = 2^\alpha \text{ for } 0 < \eta \leq \alpha^+ .$$

Let $B = A_{\alpha^+}$. By Theorem 4.2.9, $A \prec B$. Let $b \in S^\alpha$. Then for some $\eta < \alpha^+$, $b \in (S_\eta)^\alpha$. By an easy induction based on Corollary 4.2.10, we have for each ordinal ν such that $\eta < \nu \leq \alpha^+$,

$$\mathrm{Th}^+((A_\nu, b)) = \overline{\mathrm{Th}^+((A_\nu, b))} = \overline{\mathrm{Th}^+((A_\eta, b))} .$$

Taking $\nu = \alpha^+$, we see that $\mathrm{Th}^+((B, b))$ is closed. The theorem is proved.

We remark that for those readers who, for one reason or another, were unable to prove Exercise 1H, Theorem 6.1.9 can still be proved. Instead of Exercise 1H, use Exercise 5H, Theorem 1.5.10, and a chain of models of length 2^α.

EXERCISE 6E. Suppose $\|\mathscr{L}\| \leq \alpha$, $A' \prec B'$, $\omega \leq \|A'\| \leq 2^\alpha \leq \|B'\|$. Then there exists a pair of models A, B such that $A' \prec A \prec B$, $A' \prec B' \prec B$, $\|A\| = 2^\alpha$, $\|B\| = \|B'\|$, and for every $a \in R'^\alpha$,

$$\mathrm{Th}^+((A, a)) = \overline{\mathrm{Th}^+((A, a))} = \overline{\mathrm{Th}^+((A', a))} .$$

(Hint: Exercise 5J.)

EXERCISE 6F. Suppose $\|\mathscr{L}\| \leq \alpha$, $A' \prec B'$, $\omega \leq \|A'\| \leq 2^\alpha \leq \|B'\|$. Then there exists a pair of models A, B such that $A' \prec A \prec B$, $A' \prec B' \prec B$, $\|A\| = 2^\alpha$, $\|B\| = \|B'\|$, and A is α^+-saturated.

EXERCISE 6G. A is said to be *weakly α-saturated* if for every $\nu < \alpha$, every $a \in R^\nu$, and every set $\Psi \subset \Sigma_{\mathscr{L}(\nu)}{}^+$, such that $|\Psi| < \alpha$, the set $\mathrm{Th}^+((A, a)) \restriction \Psi$ is closed in X^Ψ. Using this notion the following results may be sharpened by replacing everywhere the cardinal $\|\mathscr{L}\|$ by the cardinal $|\mathscr{S}_0| \cup \omega$ and α-saturated by weakly α-saturated: Theorem 5.5.4, Corollary 5.5.5, Corollary 5.5.6, Theorem 6.1.9, and Exercise 6F.

6.2. Existence of special models

A model A is said to be *special* (with respect to $\mathscr{L}$) if A is the union of an elementary chain A_β, $\beta < \|A\|$, where each A_β is β^+-saturated. The chain A_β, $\beta < \|A\|$, is said to be a *specializing chain* for A. Note that a specializing chain is indexed by cardinals $\beta < \|A\|$.

LEMMA 6.2.1. Every α-saturated model of power α is special.

PROOF. The chain $A_\beta = A$, $\beta < \alpha$, is a specializing chain for A.

COROLLARY 6.2.2. Every finite model is special.

COROLLARY 6.2.3. A model of power α^+ is special if and only if it is α^+-saturated.

PROOF. Let A be a model of power α^+. If A is special, then A is the union of a specializing chain A_β, $\beta < \alpha^+$. Hence $A = A_\alpha$ and A is α^+-saturated. The converse follows from Lemma 6.2.1.

COROLLARY 6.2.4. Assume the generalized continuum hypothesis. Special models exist in each power α^+.

PROOF By Theorem 6.1.9 (or Corollary 5.5.6) and Corollary 6.2.3.

EXERCISE 6H*. (i) In an appropriate logic $\mathcal{L}$, find a model of power ω which is special but is not ω-saturated. (ii) Assuming that $\mathcal{L}$ has a t-set, k-set, and e-set, and that $cf(\alpha) = \alpha$, show that every special model of power α is α-saturated. This result can be applied with $\alpha = \omega$.

We recall that in Chapter I we define α^* to be the cardinal sum $\Sigma_{\beta < \alpha} 2^\beta$. We are particularly interested in those cardinals α such that $\alpha = \alpha^*$.

6.2.5. $\alpha = \alpha^*$ if and only if $2^\beta \leq \alpha$ for every $\beta < \alpha$.

6.2.6. There are arbitrarily large cardinals α such that $\alpha = \alpha^*$.

To see this let β be any cardinal. We define $\beta_0 = \beta$ and, for each $n < \omega$, $\beta_{n+1} = 2^{\beta_n}$. Let $\alpha = \Sigma_{n<\omega} \beta_n$.

6.2.7. $\alpha^+ = (\alpha^+)^*$ if and only if $\alpha^+ = 2^\alpha$.

6.2.8. The generalized continuum hypothesis is equivalent to the statement that $\alpha = \alpha^*$ for all infinite α.

We now prove the main existence theorem for special models.

THEOREM 6.2.9. Suppose $\alpha = \alpha^*$, $\|\mathcal{L}\| < \alpha$, and $\omega \leq \|A\| < \alpha$. Then there is a special model B of power α such that $A \prec B$.

PROOF. If $\alpha = \gamma^+$, then by 6.2.7 $\gamma^+ = 2^\gamma$. Hence, the theorem follows from Theorem 6.1.9 and Corollary 6.2.3. Therefore let us assume that α is a limit cardinal. Notice that if $\beta < \alpha$, then $\beta^+ < \alpha$ and $2^\beta \leq \alpha$. Let $\|\mathcal{L}\| \cup \|A\| = \gamma$. We construct an elementary chain of models B_β, $\gamma \leq \beta < \alpha$, as follows: (we use Theorem 6.1.9 and Theorem 4.2.9). B_γ is any γ^+-saturated elementary extension of A of power 2^γ; B_{β^+} is any β^{++}-saturated elementary extension of B_β of power 2^{β^+}; if β is a limit cardinal less than α, B_β is any β^+-saturated elementary extension of $\bigcup_{\delta < \beta} B_\delta$ of power 2^β.

Notice that in our construction the power of each B_β is exactly 2^β. Let $B = \bigcup_{\gamma \leq \beta < \alpha} B_\beta$. It is clear that $A \quad B$ and $\|B\| = \alpha$. Now for $\beta \leq \gamma$, define $B_\beta = B_\gamma$. The elementary chain B_β, $\beta < \alpha$, is a specializing chain for B. The theorem is proved.

The next exercise is an improvement of Theorem 6.2.9.

EXERCISE 6I. Suppose that $\alpha = \alpha^*$, $\|\mathcal{L}\| < \alpha$, and $\omega \leq \|A\| \leq \alpha$. Then there exists a special model B of power α such that $A \prec B$. (Hint: Exercise 6F.)

In the construction of B given in Theorem 6.2.9, the specializing chain B_β, $\beta < \alpha$, has the property that for all $\beta \geq \|\mathcal{L}\| \cup \|A\|$, B_β is of power 2^β. This motivates the following two exercises.

EXERCISE 6J. Suppose A is special of power α. Then A has a specializing chain A_β, $\beta < \alpha$, such that whenever $\|\mathcal{L}\| \leq \beta < \alpha$, $\|A_\beta\| \leq 2^\beta$. (Hint: Exercise 6D.)

EXERCISE 6K. Suppose $\|\mathcal{L}\| < \alpha$, $\alpha = \alpha^* = \|A\| = \|B\|$, $A \prec B$, and A is special. Then there are a special elementary extension B' of B of power α and two specializing chains A_β, $\beta < \alpha$, and B'_β, $\beta < \alpha$, for A and B', respectively, such that $A_\beta \prec B'_\beta$ for each $\beta < \alpha$.

EXERCISE 6L. A is *weakly special* if A is the union of an elementary chain A_β, $\beta < \|A\|$, where each A_β is weakly β^+-saturated. (For this notion see Exercise 6G.) Prove that Theorem 6.2.9 and Exercise 6I may be sharpened by replacing the cardinal $\|\mathcal{L}\|$ by $|\mathcal{S}_0| \cup \omega$ and special by weakly special.

6.3. Universal models

For the remainder of Chapter VI—§6.3, §6.4 and §6.5—we shall assume that $\mathcal{L}$ has a t-set, k-set, and e-set. It turns out that while the existence of special models in certain infinite powers $\alpha = \alpha^*$ does not require any special properties of $\mathcal{L}$ (and the existence of special models in each power $\alpha > \|\mathcal{L}\|$ requires the generalized continuum hypothesis), we can not prove that these special models all have the desired properties built into them unless $\mathcal{L}$ has a t-set, k-set, and e-set. This seems very strange, in particular, as regards the uniqueness of special models (Theorem 6.4.1). The reason is that the equality $\mathrm{Th}(A) = \mathrm{Th}(B)$ in an arbitrary logic $\mathcal{L}$ is too weak to have any consequences which will affect the structures of A and B. Notice that almost all of our results so far for arbitrary $\mathcal{L}$ are restricted to studying either elementary submodels of a given model A or elementary extensions of A. The following exercise is a good illustration.

EXERCISE 6M. In a logic which does not have a t-set, k-set, and e-set give an example of two finite models A, B such that $\mathrm{Th}(A) = \mathrm{Th}(B)$, $\|A\| = \|B\|$, but not $A \cong B$.

Let Θ be a theory. A model B is said to be *α-universal with respect to* Θ if every model $A \in \mathrm{Mod}(\Theta)$ of power at most α can be isomorphically embedded in B. We are concerned with the problem of finding α-universal models B such that $\|B\| = \alpha$ and $B \in \mathrm{Mod}(\Theta)$. We say that a model B is *α-universal* if it is α-universal with respect to Th(B). We say that a model B is *elementarily α-universal* if every model A such that $\mathrm{Th}(A) = \mathrm{Th}(B)$ and $\|A\| \leq \alpha$ can be elementarily embedded in B.

LEMMA 6.3.1. Suppose that $\mathrm{Th}(A) = \mathrm{Th}(B)$, B is 1-saturated, and $a_0 \in R$. Then there exists a $b_0 \in S$ such that

$$[A, a_0] \equiv [B, b_0]$$

PROOF. Let $h = [A, a_0]$. Hence $h \in \mathrm{Th}^+(A)$. By hypothesis and Theorem 4.1.5,

$$\overline{\mathrm{Th}^+(A)} = \overline{\mathrm{Th}^+(B)} = \mathrm{Th}^+(B) .$$

Hence $h \in Th^+(B)$. This means there exists an element $b_0 \in S$ such that $h = [B, b_0]$.

COROLLARY 6.3.2. Suppose that $Th(A) = Th(B)$, B is 1-saturated, and $a_0 \in R$. Then there exists a $b_0 \in S$ such that

$$Th((A, a_0)) = Th((B, b_0)) \quad .$$

PROOF. This is immediate from Lemma 6.3.1 and the simple Exercises 3J-3M on substitution.

LEMMA 6.3.3. Suppose that $Th(A) = Th(B)$, B is α-saturated, and $a \in R^\alpha$. Then there exists a sequence $b \in S^\alpha$ such that

$$Th((A, a)) = Th((B, b)) \quad .$$

PROOF. We shall construct by transfinite induction a sequence $b \in S^\alpha$ such that for all $\nu \leq \alpha$,

(1) $$Th((A, a \restriction \nu)) = Th((B, b \restriction \nu)).$$

Suppose $\nu < \alpha$ and we have already chosen b_μ, $\mu < \nu$, so that (1) holds. By 6.1.4, $(B, b \restriction \nu)$ is 1-saturated with respect to $\mathcal{L}(\nu)$, which still has a t-set, k-set, and e-set. So by Corollary 6.3.2 applied to the logic $\mathcal{L}(\nu)$, we can find an element $b_\nu \in S$ such that

$$Th(((A, a \restriction \nu), a_\nu)) = Th(((B, b \restriction \nu), b_\nu)) \quad .$$

This immdeiately implies, by Exercise 3M,

$$Th((A, a \restriction (\nu+1))) = Th((B, b \restriction (\nu+1))) \quad .$$

So (1) holds for $\nu + 1$. Suppose η is a limit ordinal such that $0 < \eta \leq \alpha$ and (1) holds for all $\nu < \eta$. Then using Exercise 3M and the fact that (see Exercise 2A)

$$\Sigma_{\mathcal{L}(\eta)} = \bigcup_{\nu < \eta} \Sigma_{\mathcal{L}(\nu)} \quad ,$$

we easily obtain (1) for η. This completes the induction. Setting $\nu = \alpha$ gives us the conclusion of the lemma.

THEOREM 6.3.4. If B is α-saturated, then B is elementarily α-universal.

PROOF. Suppose $Th(A) = Th(B)$ and $\|A\| \leq \alpha$. Let $a \in R^{\alpha}$ be an enumeration of R. By Lemma 6.3.3, there is a $b \in S^{\alpha}$ such that

$$Th((A, a)) = Th((B, b)) \quad .$$

By Theorem 4.2.8, this means A is elementarily embeddable in B.

COROLLARY 6.3.5. If B is a finite model and $Th(A) = Th(B)$, then $A \cong B$.

PROOF. By Theorems 6.1.7 and 6.3.4, B is α-universal for every α. Hence, A is isomorphic to a submodel $B' \subset B$. Thus $\|A\| \leq \|B\|$, so A is also finite. By the same argument, B is isomorphically embeddable in A, so $\|B\| \leq \|A\|$. This means $\|B'\| = \|B\|$ and $B' = B$. Therefore, $A \cong B$.

COROLLARY 6.3.6. Every complete theory with no infinite models has, up to isomorphism, exactly one model.

EXERCISE 6N. Suppose $\|\mathcal{L}\| \leq \alpha$. Then A is α-saturated if and only if for all $\nu < \alpha$ and all $a \in R^{\nu}$,

$$Th^{\mu}((A, a)) = \overline{Th^{\mu}((A, a))} \quad \text{for all} \quad \mu \leq \alpha \ .$$

For the definition of Th^{μ}, see Exercise 4A. (Hint: Use the compactness theorem for the logic $\mathcal{L}(\nu)(\mu)$, the downward Löwenheim-Skolem theorem, and Theorem 6.3.4.)

THEOREM 6.3.7. If B is a special model of power α, then B is elementarily α-universal.

PROOF. Suppose $Th(A) = Th(B)$ and $\|A\| \leq \alpha$. Let B_{β}, $\beta < \alpha$, be a specializing chain for B. Let $a \in R^{\alpha}$ be an enumeration of R. We construct by transfinite induction a sequence $b \in S^{\alpha}$ such that for each $\beta < \alpha$,

$$(1) \quad b \restriction \beta^{+} \in (S_{\beta})^{\beta^{+}} \quad \text{and} \quad Th((A, a \restriction \beta^{+})) = Th((B, b \restriction \beta^{+})) \quad .$$

Let $\gamma < \alpha$ and suppose (1) holds for every $\beta < \gamma$. Then we have

$$b \restriction \gamma \in (S_{\gamma})^{\gamma} \quad ,$$

and since $B_{\gamma} \prec B$,

$$\mathrm{Th}((A, a\restriction \gamma)) = \mathrm{Th}((B_\gamma, b\restriction \gamma)) \quad .$$

Since B_γ is γ^+-saturated, by 6.1.4, $(B_\gamma, b\restriction \gamma)$ is again γ^+-saturated with respect to $\mathcal{L}(\gamma)$. Applying Lemma 6.3.3 in the logic $\mathcal{L}(\gamma)$, we can extend the construction to a sequence $b \in (S_\gamma)^{\gamma^+}$ so that

$$\mathrm{Th}((A, a\restriction \gamma^+)) = \mathrm{Th}((B_\gamma, b\restriction \gamma^+)) \quad .$$

Hence (1) holds for γ. Our induction is complete. Now, independently of whether α is a limit cardinal or not, we obtain from (1)

$$\mathrm{Th}((A, a)) = \mathrm{Th}((B, b)) \quad .$$

By Theorem 4.2.8, A can be elementarily embedded in B. •

COROLLARY 6.3.8. Let Θ be a complete theory with infinite models. Then in every power α such that $\|\mathcal{L}\| < \alpha$ and $\alpha = \alpha^*$ there exists an elementarily α-universal model of Θ with power α.

PROOF. Using the Löwenheim-Skolem theorem and Theorem 6.2.9, there is a special model of Θ of power α. The conclusion follows from Theorem 6.3.7.

COROLLARY 6.3.9. The following are equivalent:

(i) $\mathrm{Th}(A) = \mathrm{Th}(B)$.

(ii) There exists a model A' such that both A and B can be elementarily embedded in A'.

PROOF. The proof of (i) from (ii) is trivial. So assume (i). If A is finite, then by Corollary 6.3.5, $A \cong B$. Hence (ii) follows. If A is infinite, then by applying Corollary 6.3.8 to Th(A), we again have (ii).

The following is a generalization of Corollaries 6.3.8 and 6.3.9 to theories which may not be complete.

THEOREM 6.3.10. Let Θ be a consistent theory. Then the following are equivalent:

(i) In every power α such that $\|\mathcal{L}\| \cup |X| < \alpha$

and $\alpha = \alpha^*$, Θ has an α-universal model with power α.

(ii) Any two models of Θ can be isomorphically embedded in a third model of Θ.

PROOF. We need only concern ourselves with the proof of (i) from (ii). Let α be such that $\|\mathcal{L}\| \cup |X| < \alpha$ and $\alpha = \alpha^*$. There are at most α complete theories Θ' such that $\Theta' \subset \Theta$. Corresponding to each such complete theory Θ' we choose a special model $A_{\Theta'}$ of power α such that $\mathrm{Th}(A_{\Theta'}) = \Theta'$. Using the compactness theorem in the logic $\mathcal{L}(\alpha)$ and the downward Löwenheim-Skolem theorem, we can find a model $A \in \mathrm{Mod}(\Theta)$ such that every $A_{\Theta'}$ is isomorphically embeddable in A and such that $\|A\| = \alpha$. (The details of this follow from Exercise 5I.) Now, every model of Θ is a model for some Θ'. Hence (i) follows from the fact that each $A_{\Theta'}$ was picked to be α-universal.

EXERCISE 6O. Using both parts of Exercise 5I, show that Theorem 6.3.10 can be improved by weakening condition (ii).

6.4. Uniqueness of special models

A slight alteration of the argument proving Theorem 6.3.7 yields the following:

THEOREM 6.4.1. Let A and B be special models. If $\mathrm{Th}(A) = \mathrm{Th}(B)$ and $\|A\| = \|B\|$, then $A \cong B$.

PROOF. Let $\alpha = \|A\|$ and let A_β, $\beta < \alpha$, and B_β, $\beta < \alpha$, be specializing chains for A, B, respectively. If α is finite, then by Corollary 6.3.5 $A \cong B$. So, let us assume α is infinite. Now, independent of whether α is a limit cardinal or not, we may choose enumerations $a \in R^\alpha$ of R and $b \in S^\alpha$ of S such that for all $\nu < \alpha$ we have

$$a_\nu \in R_{|\nu|} \quad \text{and} \quad b_\nu \in S_{|\nu|} \quad .$$

In what follows we shall reserve the letter η as a variable which ranges over limit ordinals (including 0). We shall construct by transfinite induction two sequences $c \in R^\alpha$ and $d \in S^\alpha$ such that for each $\nu < \alpha$

with $|\nu| = \gamma$,

(1) $$c_\nu \in R_\gamma \text{ and } d_\nu \in S_\gamma \quad ;$$

(2) $$\mathrm{Th}((A, c \upharpoonright (\nu+1)) = \mathrm{Th}((B, d \upharpoonright (\nu+1))) \quad ;$$

(3) $$\text{if } \nu = \eta + 2n, \text{ then } c_\nu = a_{\eta+n}, \text{ and}$$
$$\text{if } \nu = \eta + 2n + 1, \text{ then } d_\nu = b_{\eta+n} \quad .$$

Suppose $\mu < \alpha$, $|\mu| = \beta$, and (1), (2), (3) hold for all $\nu < \mu$. Then by (2), we have

(4) $$\mathrm{Th}((A, c \upharpoonright \mu)) = \mathrm{Th}((B, d \upharpoonright \mu)) \quad ,$$

and by (1)

(5) $$c \upharpoonright \mu \in R^\mu_\beta \text{ and } d \upharpoonright \mu \in S^\mu_\beta \quad .$$

Since $A_\beta \prec A$ and $B_\beta \prec B$, we have from (4) and (5),

$$\mathrm{Th}((A_\beta, c \upharpoonright \mu)) = \mathrm{Th}((B_\beta, d \upharpoonright \mu)) \quad .$$

By Theorem 4.1.5 applied to the logic $\mathcal{L}(\mu)$, we have

$$\overline{\mathrm{Th}^+((A_\beta, c \upharpoonright \mu))} = \overline{\mathrm{Th}^+((B_\beta, d \upharpoonright \mu))} \quad .$$

Since both A_β and B_β are β^+-saturated and since $\mu < \beta^+$, the above yields

(6) $$\mathrm{Th}^+((A_\beta, c \upharpoonright \mu)) = \mathrm{Th}^+((B_\beta, d \upharpoonright \mu)).$$

If $\mu = \eta + 2n$, define $c_\mu = a_{\eta+n}$, then by (6) we may choose $d_\mu \in S_\beta$ such that

(7) $$[(A_\beta, c \upharpoonright \mu), c_\mu] = [(B_\beta, d \upharpoonright \mu), d_\mu] \quad .$$

On the other hand, if $\mu = \eta + 2n + 1$, define $d_\mu = b_{\eta+n}$, and using (6) again we may pick $c_\mu \in R_\beta$ so that (7) again holds. In either case (7) leads immediately to

$$\mathrm{Th}((A, c \upharpoonright (\mu+1))) = \mathrm{Th}((B, d \upharpoonright (\mu+1))) \quad ,$$

and conditions (1), (2), and (3) hold for μ. Thus our induction is complete and we may assume that c and d have been constructed. It is now evident that c and d enumerate R and S respectively and that

$$\mathrm{Th}((A, c)) = \mathrm{Th}((B, d)) \quad .$$

Hence, by Exercise 3G, $A \cong B$.

EXERCISE 6P. A class $K \subset \mathcal{M}$ is elementarily closed if and only if both K and $\mathcal{M} - K$ are

(i) closed under isomorphisms;

(ii) closed under unions of elementary chains;

(iii) closed under ultrapowers.

EXERCISE 6Q. $K \in EC_\Delta$ iff K and $\mathcal{M} - K$ have properties (i) and (ii) of Exercise 6P, K is closed under ultraproducts, and $M - K$ is closed under ultrapowers.

EXERCISE 6R*. Suppose that $B_1 \subset A$, $B_2 \subset A$, and $Th(B_1) = Th(B_2)$. Then there exist three models B_1', B_2' and A' such that

$$B_1' \cong B_2', \quad A \prec A', \quad B_1 \prec B_1' \subset A', \quad B_2 \prec B_2' \subset A' \quad .$$

(Hint: See Exercise 6F and try to generalize.) We remark that the result of this exercise for the logic ℓ may be stated in a somewhat stronger form.

EXERCISE 6S*. For subsets $z \subset \pi$, let Σ_z denote the set of all sentences of Σ in which no predicates P_ξ, $\xi \notin z$, occur. As usual, let $\Theta \restriction \Sigma_z$ denote the restriction of Θ to the space X^{Σ_z}. We say that $\Theta \restriction \Sigma_z$ is consistent (complete) if there exists a model A such that $[A] \restriction \Sigma_z \in \Theta \restriction \Sigma_z$ $(Th(A) \restriction \Sigma_z = \Theta \restriction \Sigma_z)$. Suppose that z_1, $z_2 \subset \pi$ and $\Theta \restriction \Sigma_{z_1 \cap z_2}$ is complete. Then the following are equivalent:

(i) $\Theta \restriction \Sigma_{z_1}$ and $\Theta \restriction \Sigma_{z_2}$ are both consistent;

(ii) $\Theta \restriction \Sigma_{z_1 \cup z_2}$ is consistent.

(Hint: A model which is special with respect to $\mathcal{L}$ is also special with respect to that part of $\mathcal{L}$ which contains only those predicates P_ξ, $\xi \in z$.)

EXERCISE 6T*. Following the notation of the last exercise, we say that a class $K \subset \mathcal{M}$ is a z-class if $A \in K$ and $[B] \restriction \Sigma_z = [A] \restriction \Sigma_z$ imply $B \in K$. Suppose $K_1, K_2 \in EC_\Delta$, K_1 is a z_1-class and K_2 is a z_2-class. Then the following are equivalent:

(i) $K_1 \cap K_2 = 0$;

(ii) there are $(z_1 \cap z_2)$-classes $L_1, L_2 \in EC_\Delta$ such that $K_1 \subset L_1$, $K_2 \subset L_2$, and $L_1 \cap L_2 = 0$.

The following theorem is a certain completeness result for every continuous logic $\mathcal{L}$ which has a t-set, k-set, and e-set, and it justifies the importance we have attached to these sets.

THEOREM 6.4.2. The following are equivalent.

(i) A and B are elementarily equivalent with respect to $\mathcal{L}$.

(ii) A and B are elementarily equivalent with respect to the logic $\mathcal{L}(\mathcal{C} \cup \mathcal{Q})$.

(iii) A and B are elementarily equivalent with respect to every logic $\mathcal{L}(\mathcal{F}')$ where $\mathcal{F}' \subset \mathcal{C} \cup \mathcal{Q}$.

PROOF. It is sufficient to prove (ii) from (i), as the other implications are all obvious. Suppose $[A]_{\mathcal{L}} = [B]_{\mathcal{L}}$. If A is finite, then $A \cong B$ by Corollary 6.3.5. Since in the logic $\mathcal{L}(\mathcal{C} \cup \mathcal{Q})$ we have not added any new predicate or individual constant symbols, we have, in this case $[A]_{\mathcal{L}(\mathcal{C} \cup \mathcal{Q})} = [B]_{\mathcal{L}(\mathcal{C} \cup \mathcal{Q})}$. So, let A be infinite. Then B is also infinite. Let $\alpha = \alpha^*$ be a cardinal such that

$$\|A\| \cup \|B\| \cup \|\mathcal{L}(\mathcal{C} \cup \mathcal{Q})\| < \alpha \ .$$

Apply Theorem 6.2.9 with the logic $\mathcal{L}(\mathcal{C} \cup \mathcal{Q})$ and find two special models A' and B' of power α such that

$$A \prec A' \text{ and } [A]_{\mathcal{L}(\mathcal{C} \cup \mathcal{Q})} = [A']_{\mathcal{L}(\mathcal{C} \cup \mathcal{Q})} \ ,$$

$$B \prec B' \text{ and } [B]_{\mathcal{L}(\mathcal{C} \cup \mathcal{Q})} = [B']_{\mathcal{L}(\mathcal{C} \cup \mathcal{Q})} \ .$$

Since $[A]_{\mathcal{L}} = [B]_{\mathcal{L}}$, we have $[A']_{\mathcal{L}} = [B']_{\mathcal{L}}$. By Theorem 6.4.1, $A' \cong B'$. Hence

$$[A']_{\mathcal{L}(\mathcal{C} \cup \mathcal{Q})} = [B']_{\mathcal{L}(\mathcal{C} \cup \mathcal{Q})} \ .$$

This means $[A]_{\mathcal{L}(\mathcal{C} \cup \mathcal{Q})} = [B]_{\mathcal{L}(\mathcal{C} \cup \mathcal{Q})}$ and the theorem is proved.

COROLLARY 6.4.3. Suppose the logics $\mathcal{L}$ and $\mathcal{L}'$ differ only in that $\mathcal{F} \neq \mathcal{F}'$, and suppose both $\mathcal{L}$ and $\mathcal{L}'$ have a t-set, k-set, and e-set. Then

$$[A]_{\mathcal{L}} = [B]_{\mathcal{L}} \text{ if and only if } [A]_{\mathcal{L}'} = [B]_{\mathcal{L}'} .$$

EXERCISE 6U. Under the same hypotheses as in Corollary 6.4.3, show that

$$\overline{Th^+(A)}_{\mathcal{L}} = \overline{Th^+(B)}_{\mathcal{L}} \text{ if and only if } \overline{Th^+(A)}_{\mathcal{L}'} = \overline{Th^+(B)}_{\mathcal{L}'} .$$

EXERCISE 6V*. Let X' be a compact Hausdorff space which has X as a subspace. Let

$$\mathcal{L}' = (\langle\pi, \kappa\rangle, X', \mathcal{F}', \underset{\sim}{0}, \underset{\sim}{1}, H')$$

be a continuous logic. Thus $\mathcal{M}_{\mathcal{L}} \subset \mathcal{M}_{\mathcal{L}'}$. Show that if $A, B \in \mathcal{M}_{\mathcal{L}}$ and $[A]_{\mathcal{L}} = [B]_{\mathcal{L}}$, then $[A]_{\mathcal{L}'} = [B]_{\mathcal{L}'}$. (Hint: Look at the detailed construction of special models and at the various remarks in §5.1 concerning the independence of the ultraproduct construction from the logic $\mathcal{L}$.)

6.5. Some consequences of the generalized continuum hypothesis

In this section we combine our results of Sections 5.4 and 5.5 on α-good ultrafilters with the uniqueness theorem for special models. We assume the generalized continuum hypothesis throughout this section, and also continue to assume that $\mathcal{L}$ has a t-set, k-set and e-set.

THEOREM 6.5.1. For A and B to be elementarily equivalent it is necessary and sufficient that there exist a set I and an ultrafilter D over I such that

$$D\text{-prod } A \cong D\text{-prod } B .$$

Moreover, if $\|\mathcal{L}\| \leq \alpha$ and $\|A\|, \|B\| \leq 2^\alpha$, then we may take I to be of power α.

PROOF. The sufficiency follows, by Corollary 5.1.7, from the equations

$$Th(A) = Th(D\text{-prod } A) = Th(D\text{-prod } B) = Th(B) .$$

To prove necessity, let α be sufficiently large so that $\|\mathcal{L}\| \leq \alpha$ and $\|A\|, \|B\| \leq 2^\alpha$. As in the proof of Corollary 5.5.6, we note that there exists an α^+-good, countably incomplete ultrafilter D over α, and that the ultrapowers D-prod A and D-prod B are α^+-saturated models of power $\leq \alpha^+$. Applying 5.1.7 we have

$$\mathrm{Th}(\text{D-prod } A) = \mathrm{Th}(A) = \mathrm{Th}(B). = \mathrm{Th}(\text{D-prod } B) \quad .$$

If A is finite, then by 6.3.5 we have

$$\text{D-prod } A \cong A \cong B \cong \text{D-prod } B \quad .$$

If A is infinite, then D-prod A and D-prod B are infinite and, by Lemma 6.1.6, D-prod A and D-prod B are of power exactly α^+. Then by Lemma 6.2.1, D-prod A and D-prod B are special models. Finally, by Theorem 6.4.1, we have

$$\text{D-prod } A \cong \text{D-prod } B \quad .$$

Notice that in the above proof the generalized continuum hypothesis was used twice: first to show the existence of the α^+-good ultrafilter D, and second to show that the D-products are of power $\leq \alpha^+$. In case $\alpha = \omega$, only the second use of the continuum hypothesis is needed.

In the following discussion we assume that $K \subset \mathcal{M}$ and that K is closed under isomorphisms. We say that K is *closed under ultrapowers* if D-prod B $\in$ K whenever B $\in$ K, I is a nonempty set, and D is an ultrafilter over I.

COROLLARY 6.5.2. For K to be elementarily closed it is necessary and sufficient that both K and $\mathcal{M} - K$ be closed under ultrapowers.

PROOF. This is an easy consequence of Theorem 6.5.1.

COROLLARY 6.5.3. For K to be an elementary class it is neccessary and sufficient that K be closed under ultraproducts and $\mathcal{M} - K$ be closed under ultrapowers.

PROOF. By 5.2.4 and 6.5.2.

THEOREM 6.5.4. Let $K, L \subset \mathcal{M}$. For

$$\overline{\mathrm{Th}(K)} \cap \overline{\mathrm{Th}(L)} \neq 0$$

it is necessary and sufficient that there exist a set I, an ultrafilter D over I, and functions $F \in K^I$, $G \in L^I$ such that $D\text{-prod } F \cong D\text{-prod } G$.

PROOF. The sufficiency follows from 5.1.6 and 1.5.2.

To prove the necessity, let $h \in \overline{Th(K)} \cap \overline{Th(L)}$. Let T be the set of all basic open neighborhoods of h in the space X^Σ. Then for each $U \in T$, there exists $A_U \in K$, $B_U \in L$ such that $[A_U], [B_U] \in U$. Let β be a cardinal such that $\|A_U\| \le 2^\beta$ and $\|B_U\| \le 2^\beta$ for all $U \in T$, let $I = S_\omega(T \times \beta)$, and let $\alpha = |I|$. For each

$$i = \{\langle U_1, \zeta_1\rangle, \ldots, \langle U_n, \zeta_n\rangle\} \in I \quad ,$$

let $A_i = A_U$ and $B_i = B_U$, where $U = U_1 \cap \ldots \cap U_n$. By Exercise 5O there is an α^+-good, regular ultrafilter D over I. By Exercise 1J, D is countably incomplete. Hence by Theorem 5.5.4, each of the models

$$A' = D\text{-prod } \lambda i\, A_i, \quad B' = D\text{-prod } \lambda i\, B_i$$

is α^+-saturated. Moreover, we have

$$\|A'\| \le |\prod_{i \in I} R_i| \le (2^\beta)^\alpha = \alpha^+ ,$$

and similarly $\|B'\| \le \alpha^+$. Since D is regular, we have for each $U \in T$,

$$\{i \in I: \langle U, 0\rangle \in I\} \subset \{i \in I: [A_i] \in U\} \in D \quad ,$$

and therefore

$$D\text{-lim } \lambda i[A_i] \in U \quad .$$

It follows, using 5.1.6, that

$$[A'] = D\text{-lim } \lambda i[A_i] = h \quad .$$

By a similar argument we have $[B'] = h$, and hence $Th(A') = Th(B')$. Arguing as in the proof of Theorem 6.5.1, we see that $A' \cong B'$, and our proof is complete.

EXERCISE 6W (Separation theorem). Suppose $K, L \subset \mathcal{M}$, $K \cap L = 0$, and K and L are both closed under ultraproducts. Then there exist $K', L' \subset \mathcal{M}$ such that $K \subset K'$, $L \subset L'$, $K' \cap L' = 0$, and K', L' are both intersections of singular elementary classes.

EXERCISE 6X. (Compare with Exercise 5F.) For both K and $\mathcal{M} - K$ to be finite intersections of singular elementary classes it is necessary and sufficient that both K and $\mathcal{M} - K$ are closed under ultraproducts.

CHAPTER VII

CLASSES PRESERVED UNDER ALGEBRAIC RELATIONS

In Sections 7.2 and 7.3 of this chapter we shall generalize two well-known results in the two-valued logic ℓ to a large class of continuous logics $\mathscr{L}$. The known results in ℓ are: (1) An elementary class K is closed under extensions of models if and only if K is the class of all models of a set of existential sentences (Łoś [1955b] and Tarski [1954]). (2) An elementary class K is closed under homomorphic images if and only if K is the class of all models of a set of positive sentences (Lyndon [1959]). These are but two of a long series of results in ℓ which say that an elementary class K is closed under some relation if and only if K is the class of all models of a set of sentences of a certain kind. We shall indicate in the exercises of these sections how one can generalize some of the other results in that series to the case of continuous model theory.

Section 7.4 of this chapter has a more special character in that it deals with continuous logics $\mathscr{L}$ in which the relation H is a simple order relation over X. We are led to such restricted logics $\mathscr{L}$ because we are apparently unable to define a meaningful reduced product otherwise. The result in two-valued logic which we generalize is the following: (3) An elementary class K is closed under reduced products if and only if K is the class of all models of a set of conditional sentences (Keisler [1965]). In contrast with (1) and (2), this result requires the generalized continuum hypothesis and, to date, it is not known if a proof can be found without using the generalized continuum hypothesis.

We recall that the order relation, H, of $\mathscr{L}$ has not been mentioned at all in the last three chapters (IV, V, and VI). In this chapter, however, the relation H is vital and we shall make essential use of our assumption from Chapter II that $(X, \underset{\sim}{0}, \underset{\sim}{1})$ is ordered by H in any continuous logic $\mathcal{L}$.

7.1. The extended theory and order

In this section we prepare the way for the rest of the chapter by generalizing the results of Section 4.1. Specifically, we shall repeat the discussion in 4.1 but with an H-set $\mathcal{H}$ in place of the t-set. Throughout this section we make the following assumptions on $\mathcal{L}$:

7.1.1. $\mathcal{L}$ has an H-set $\mathcal{H}$;

7.1.2. $\mathcal{L}$ has a k-set $\mathcal{K}$;

7.1.3. $\mathcal{L}$ has an e-set $\mathcal{E}$.

Notice that, by Lemma 1.4.1, it follows from 7.1.1 that the relations H and $\breve{H}$ are closed with respect to X.

As in Section 4.1, we shall let Ψ be an arbitrary subset of Σ^+. We shall depart from 4.1, however, in the following definition. For each $h \in X^\Psi$, the set $\Psi_h \subset \Sigma^+$ is defined by:

$$\Psi_h = \{t(\varphi) : \varphi \in \Psi, \quad t \in \mathcal{H}, \text{ and } \quad t(h(\varphi)) = \underset{\sim}{1}\}.$$

As before, we shall use the symbols $\underset{\sim}{0}$ and $\underset{\sim}{1}$ to denote constant functions whose ranges are $\{0\}$ and $\{1\}$ respectively. We shall engage in one further abuse of language, and use the symbol H to denote the binary relation over X^Ψ defined pointwise from the relation H over X; that is, for h, $h' \in X^\Psi$,

$$hHh' \text{ means that } h(\varphi)\, Hh'(\varphi) \text{ for all } \varphi \in \Psi.$$

We remark in passing that by Exercise 1P, the relations H and $\breve{H}$ over X^Ψ are closed and $(\underset{\sim}{X}^\Psi, \underset{\sim}{0}, \underset{\sim}{1})$ is ordered by H.

Our first lemma is the analog of Lemma 4.1.1.

LEMMA 7.1.4. If

$$h \in \overline{\breve{H}[Th^+(A) \upharpoonright \Psi]},$$

then

$$\underset{\sim}{1} \in \overline{Th^+(A) \upharpoonright \Psi_h}.$$

PROOF. It suffices to prove that whenever $\varphi_1, \dots, \varphi_n \in \Psi_h$ and $\underset{\sim}{1} \in V \in \mathcal{S}$, there exists an $r_0 \in R$ such that

(1) $$\varphi_m[A, r_0] \in H[V] \text{ for each } m \leq n.$$

We note that for each $m \leq n$ we have $\varphi_m = t_m(\psi_m)$, where $t_m(h(\psi_m)) = \underset{\sim}{1}$, $t_m \in \mathcal{H}$, and $\psi_m \in \Psi$. Since each t_m is continuous, we have

$$h(\psi_m) \in \breve{t}_m[V] \in \mathcal{S} \text{ for each } m \leq n.$$

By hypothesis, there exists $g \in \breve{H}[Th^+(A) \upharpoonright \Psi]$ such that $g(\psi_m) \in \breve{t}_m[V]$ for all $m \leq n$. Moreover, there exists $r_0 \in R$ such that for all $\psi \in \Psi$, $g(\psi)\ H\psi[A, r_0]$. This means that

$$\psi_m[A, r_0] \in H[\breve{t}_m[V]].$$

Since t_m is H-preserving, we have

$$H[\breve{t}_m[V]] \subset \breve{t}_m[H[V]],$$

and (1) follows readily.

We now prove the analog of Theorem 4.1.4.

THEOREM 7.1.5. Let $h \in X^\Psi$. The following two conditions are equivalent:

(i) $h \in \overline{\breve{H}[Th^+(A) \upharpoonright \Psi]}$;

(ii) $\varphi[A] = \underset{\sim}{1}$ for each $\varphi \in \Sigma \cap \mathcal{E}(\mathcal{K}(\Psi_h))$.

PROOF. The proof that (i) implies (ii) is the same as the first paragraph of the proof of Theorem 4.1.4, except that Lemma 7.1.4 is used in place of Lemma 4.1.1.

Assume that (i) fails. Then there exist $\varphi_1, \dots, \varphi_n \in \Psi$ and $h(\varphi_1) \in U_1 \in \mathcal{S}, \dots, h(\varphi_n) \in U_n \in \mathcal{S}$ such that

(1) $$\{\langle\varphi_1[A, r_0], \dots, \varphi_n[A, r_0]\rangle : r_0 \in R\} \subset X^n - (H[U_1] \times \dots \times H[U_n]).$$

Let us denote the right-hand set in the above inclusion by Z. Since the relation H is open, each set $H[U_m]$, $m \leq n$, is open, and therefore Z is closed in X^n. By Theorem 2.4.5, for each $m \leq n$ there is a finite subset

$$\mathcal{H}_m = \{t_{m1}, \ldots, t_{mi_m}\} \subset \mathcal{H}$$

such that

(2) $$t(h(\varphi_m)) = \underset{\sim}{1} \quad \text{for each} \quad t \in \mathcal{H}_m$$

and

(3) $$\text{if} \quad t(x) = \underset{\sim}{1} \quad \text{for all} \quad t \in \mathcal{H}_m, \quad \text{then} \quad x \in H[U_m] .$$

To complete the proof we may now simply repeat, word for word, the last paragraph of the proof of Theorem 4.1.4.

We conclude this section with a theorem analogous to Theorem 4.1.5.

THEOREM 7.1.6. Suppose that

$$\Psi = \mathcal{E}(\mathcal{K}(\mathcal{H}(\Psi))).$$

Then for any two models A and B, the following are equivalent:

(i) $\underset{\sim}{1} \in \overline{\mathrm{Th}(B) \restriction \check{h}[\underset{\sim}{1}]}$ for the unique $h \in \mathrm{Th}(A) \restriction (\Sigma \cap \Psi)$;

(ii) $\underset{\sim}{1} \in \overline{\mathrm{Th}^+(B) \restriction \check{h}[\underset{\sim}{1}]}$ for all $h \in \mathrm{Th}^+(A) \restriction \Psi$.

PROOF. (i) easily follows from (ii). So, let us assume (i) and prove (ii). Let $h \in \mathrm{Th}^+(A) \restriction \Psi$. Since H is reflexive, $h \in \check{H}[\mathrm{Th}^+(A) \restriction \Psi]$. By 7.1.5,

$$\varphi[A] = \underset{\sim}{1} \quad \text{for all} \quad \varphi \in \Sigma \cap \mathcal{E}(\mathcal{K}(\Psi_h)).$$

It follows from the inclusion $\Psi_h \subset \mathcal{H}(\Psi)$, and from our hypothesis, that $\mathcal{E}(\mathcal{K}(\Psi_h)) \subset \Psi$. Let f be the unqne element of $\mathrm{Th}(A) \restriction (\Sigma \cap \Psi)$, That is, $f = [A] \restriction (\Sigma \cap \Psi)$. Then we have

$$\Sigma \cap \mathcal{E}(\mathcal{K}(\Psi_h)) \subset \check{f}[\underset{\sim}{1}].$$

follows from (i) that

$$\underset{\sim}{1} \in \overline{\mathrm{Th}(B)\restriction \breve{f}[\underset{\sim}{1}]},$$

hence

$$\mathrm{Th}(B)\restriction \breve{f}[\underset{\sim}{1}] = \{\underset{\sim}{1}\} .$$

refore

$$\varphi[B] = \underset{\sim}{1} \quad \text{for all} \quad \varphi \in \Sigma \cap \mathscr{E}(\mathscr{K}(\Psi_h)).$$

ng 7.1.5 again,

$$h \in \overline{\breve{H}[\mathrm{Th}^+(B)\restriction \Psi]}.$$

ce

$$\underset{\sim}{1} \in \overline{\breve{H}\ [\mathrm{Th}^+(B)\restriction \breve{h}[\underset{\sim}{1}]]}.$$

ce $(X,\underset{\sim}{0},\underset{\sim}{1})$ is ordered by H, $\underset{\sim}{1}$ is a continuous fixed point of H h respect to X, that is, $\underset{\sim}{1} \in V \in \mathscr{S}$ implies that there is a $\mathscr{S}$ with $\underset{\sim}{1} \in H[U] \subset V$. Therefore each basic open neighborhood V' $\underset{\sim}{1}$ in the space $X^{\breve{h}[\underset{\sim}{1}]}$ includes an open neighborhood U' of $\underset{\sim}{1}$ in $[\underset{\sim}{1}]$ such that $H[U'] \subset V'$ (see Exercise 1P). By (1), U' meets $\mathrm{Th}^+(B)\restriction \breve{h}[\underset{\sim}{1}]]$, and hence $H[U']$ meets $\mathrm{Th}^+(B)\restriction \breve{h}[\underset{\sim}{1}]$. Then V' meets set $\mathrm{Th}^+(B)\restriction \breve{h}[\underset{\sim}{1}]$, and (ii) follows.

2. Extensions of models and existential formulas

Throughout this section we shall sssume the following:

7.2.1. $\mathscr{L}$ has a t-set $\mathscr{T}$;

7.2.2. $\mathscr{L}$ has an H-set $\mathscr{H}$;

7.2.3. $\mathscr{L}$ has a k-set $\mathscr{K} \subset \mathscr{C}_H$;

7.2.4. $\mathscr{L}$ has an e-set $\mathscr{E} \subset \mathscr{Q}_{\breve{H}}$.

sets $\mathscr{C}_H$ and $\mathscr{Q}_{\breve{H}}$ are defined in Section 1.4. A formula $\varphi \in \Phi$ said to be *existential* if

$$\varphi \in ((\mathscr{Q}_{\breve{H}} \cup \mathscr{C}_H) \cap \mathscr{F})(\mathscr{C} \cap \mathscr{F})\wedge .$$

In the literature, a formula φ in the two-valued logic ℓ is usually said to be "existential" if all of its quantifiers are existential and occur at the beginning of φ. It is easily seen that a formula of ℓ is equivalent to an existential formula in our sense if and only if it is equivalent to an existential formula in the usual sense from the literature. Thus our notion of an existential formula in $\mathcal{L}$ is, essentially, a generalization of the usual notion in ℓ.

We let Υ^+ be the set of all existential formulas in Σ^+, and let Υ be the set of all existential sentences in Σ. Similarly, we let Υ^+_μ, Υ_μ denote the sets of all existential formulas in $\Sigma^+_{\mathcal{L}(\mu)}$, $\Sigma_{\mathcal{L}(\mu)}$, respectively.

Notice that Exercise 3C concerns existential formulas, and it may be reformulated as follows:

7.2.5. If $\varphi \in \Upsilon^+$, $A \subset B$, and $r_0 \in R$, then $\varphi[A, r_0]H\varphi[B, r_0]$. If $\varphi \in \Upsilon$ and $A \subset B$, then $\varphi[A]H\varphi[B]$.

Our goal in this section is to find converses to 7.2.5.

THEOREM 7.2.6. Suppose that $\|B\|$ is finite and that for for every existential sentence φ we have

$$\varphi[A] = \underset{\sim}{1} \quad \text{implies} \quad \varphi[B] = \underset{\sim}{1}.$$

Then $\|A\| \leq \|B\|$.

PROOF. Suppose that $1 < n < \omega$ and $n \leq \|A\|$. Let $m = \frac{n(n-1)}{2}$. Choose $t \in \mathcal{T}$ such that $t(\underset{\sim}{0}) = \underset{\sim}{1}$ and $t(\underset{\sim}{1}) \neq \underset{\sim}{1}$. Since $\mathcal{L}$ has a k-set, we may apply Lemma 2.4.7 to show that there is a function $k \in \mathcal{C}_m \cap \mathcal{F}$ such that $k(\underset{\sim}{1}, \ldots, \underset{\sim}{1}) = \underset{\sim}{1}$ and $k[\underset{\sim}{1}] \subset (X - \{t(\underset{\sim}{1})\})^m$. Let φ_0 be the formula

$$\varphi_0 = k(t(v_1 \equiv v_2), \; t(v_1 \equiv v_3), \; \ldots, \; t(v_1 \equiv v_n), \; \ldots, \; t(v_{n-1} \equiv v_n)).$$

It is easily seen that φ_0 has the following three properties:

(1) φ_0 is existential;

(2) for any $C \in \mathcal{M}, \varphi_0[C, r] = \underset{\sim}{1}$ for some $r \in T^\infty$ if any only if $\|C\| \geq n$.

For any $\psi \in \Phi$, let us write

$$\mathrm{val}(\psi) = \{\psi[C, r] : C \in \mathscr{M} \text{ and } r \in T^{\infty}\} - \{\underset{\sim}{1}\}.$$

Then φ_0 also has the property

(3) $\mathrm{val}(\varphi_0)$ is a finite set.

We wish to construct a sentence which has the properties (1) and (2).

It follows from (3) that each set $Z \in \mathrm{val}(\varphi_0)^*$ is finite and thus $\underset{\sim}{1} \notin \overline{Z}$. Then there is a finite subset $\mathscr{E}_1 = \{e_1, \ldots, e_p\} \subset \mathscr{E}$ such that for all $Z \in \mathrm{val}(\varphi_0)^*$ there exists $e \in \mathscr{E}_1$ with $e(Z) \neq \underset{\sim}{1}$. It is clear that $\mathrm{val}((e_j v_1)\varphi_0)$ is finite for each $j \leq p$. Thus we may apply Lemma 2.4.7 and there exists $k_1 \in \mathscr{F} \cap \mathscr{C}_p \cap \mathscr{C}_H$ such that $k_1(\underset{\sim}{1}, \ldots, \underset{\sim}{1}) = \underset{\sim}{1}$ and

$$\breve{k}_1[1] \subset (X - (\mathrm{val}((e_1 v_1)\varphi_0) \cup \ldots \cup \mathrm{val}((e_p v_1)\varphi_0))^p.$$

Let

$$\varphi_1 = k_1((e_1 v_1)\varphi_0, \ \ldots, \ (e_p v_1)\varphi_0).$$

Our choice of $\mathscr{E}_1$ and k_1 guarantees that the formula φ_1 has the properties (1), (2), and (3). We now repeat the process, obtaining a sequence of formulas $\varphi_0, \varphi_1, \varphi_2, \ldots, \varphi_n$ such that for each $j < n$, φ_{j+1} is obtained from φ_j in the same way as φ_1 was obtained from φ_0. By induction we see that each φ_j has the properties (1), (2), and (3). Each φ_j has only the free variables $v_{j+1}, \ldots, v_n$, and hence φ_n is a sentence. Since $\|A\| \geq n$, it follows from (1) and (2) that

$$\varphi_n[A] = \underset{\sim}{1}, \quad \varphi_n[B] = \underset{\sim}{1}, \quad \text{and } \|B\| \geq n.$$

LEMMA 7.2.7. $T^+ = \mathscr{E}(\mathscr{K}(\mathscr{H}(T^+)))$.

PROOF. By 7.2.2. - 7.2.4 and the definition of T^+.

THEOREM 7.2.8. Suppose B is a special model, $\|A\| \leq \|B\|$, and for every existential sentence φ we have

$$\varphi[A] = \underset{\sim}{1} \text{ implies } \varphi[B] = \underset{\sim}{1}.$$

Then A is isomorphically embeddable in B.

PROOF. Let $\|B\| = \alpha$ and let B_β, $\beta < \alpha$, be a specializing chain for B. Since $\|A\| \leq \alpha$, there is an enumeration $a \in R^\alpha$ of R. We shall construct by transfinite induction a sequence $b \in S^\alpha$ such that, for each $\upsilon < \alpha$, we have

(1) for all $\varphi \in T_\upsilon$, $\varphi[(A, a\restriction\upsilon)] = \underset{\sim}{1}$ implies $\varphi[(B, b\restriction\upsilon)] = \underset{\sim}{1}$;

(2) $\mathscr{R}(b\restriction\upsilon) \subset S_{|\upsilon|}$.

By hypothesis, (1) and (2) hold when $\upsilon = 0$. Moreover, it is obvious that if υ is a limit ordinal and (1), (2) hold for all $\mu < \upsilon$, then (1) and (2) hold for υ.

Now let $\upsilon < \alpha$ and suppose that (1) and (2) hold. Let $\beta = |\upsilon|$. We may rewrite (1) as

(3) $\underset{\sim}{1} \in \text{Th}((B, b\restriction\upsilon))\restriction\breve{h}[\underset{\sim}{1}]$ for the unique $h \in \text{Th}((A, a\restriction\upsilon))\restriction T_\upsilon$.

By 4.2.9, $B_\beta \prec B$, so by 4.2.7 we have

$$(B_\beta, b\restriction\upsilon) \prec (B, b\restriction\upsilon).$$

Therefore (3) holds with B_β in place of B. By 7.2.7, we may apply Theorem 7.1.6 to the logic $\mathscr{L}(\upsilon)$, with T_υ^+ for Ψ, and we obtain

(4) $\underset{\sim}{1} \in \overline{\text{Th}^+((B_\beta, b\restriction\upsilon))\restriction\breve{h}[\underset{\sim}{1}]}$ for all $h \in \text{Th}^+((A, a\restriction\upsilon))\restriction T_\upsilon^+$.

Let

$$h_0 = [(A, a\restriction\upsilon), a_\upsilon]\restriction T_\upsilon^+ .$$

Then

$$h_0 \in \text{Th}^+((A, a\restriction\upsilon))\restriction T_\upsilon^+$$

and hence by (4),

$$\underset{\sim}{1} \in \overline{\text{Th}^+((B_\beta, b\restriction\upsilon))\restriction\breve{h}_0[\underset{\sim}{1}]} .$$

But since B_β is β^+-saturated and $\upsilon < \beta^+$, and by (2), the set $\text{Th}^+((B_\beta, b\restriction\upsilon))$ is closed. Therefore

$$\underset{\sim}{1} \in Th^+((B_\beta, b\restriction \upsilon))\restriction\aleph_0[\underset{\sim}{1}].$$

We may now choose b_υ so that $b_\upsilon \in S_\beta$ and, for all $\varphi \in \Upsilon^+_\upsilon$,

$$\varphi[(A, a\restriction \upsilon), a_\upsilon] = \underset{\sim}{1} \text{ implies } \varphi[(B_\beta, b\restriction \upsilon), b_\upsilon] = \underset{\sim}{1}.$$

Clearly (2) holds for $\upsilon + 1$, i.e. $\mathscr{R}(b\restriction \upsilon + 1) \subset S_\beta$. Moreover, we see from Exercises 3K and 3M applied to $\mathscr{L}(\upsilon)$ that

$$\varphi \in \Upsilon_{\upsilon+1} \text{ and } \varphi[(A, a\restriction \upsilon + 1)] = \underset{\sim}{1} \text{ implies } \varphi[(B_\beta, b\restriction \upsilon + 1)] = \underset{\sim}{1}.$$

Since $(B_\beta, b\restriction \upsilon + 1) \prec (B, b\restriction \upsilon + 1)$, (1) holds for $\upsilon + 1$. This completes our induction.

Since (1) holds for all $\upsilon < \alpha$, we have

(5) $\varphi \in \Upsilon_\alpha$ and $\varphi[A, a)] = \underset{\sim}{1}$ implies $\varphi[(B, b)] = \underset{\sim}{1}$.

Let $\psi \in \Lambda_{\mathscr{L}(\alpha)} \cap \Sigma_{\mathscr{L}(\alpha)}$ and let $x = \psi[(A, a)]$, $y = \psi[(B, b)]$. If $x \neq y$, then there exists $t \in \mathscr{T}$ such that $t(x) = \underset{\sim}{1}$ and $t(y) \neq \underset{\sim}{1}$. But this means that

(6) $t(\psi)[(A, a)] = \underset{\sim}{1}$ and $t(\psi)[(B, b)] \neq \underset{\sim}{1}$,

and since $t(\psi) \in (\mathscr{F} \cap \mathscr{C})\ \Lambda_{\mathscr{L}(\alpha)} \subset \Upsilon_\alpha$, (6) contradicts (5). Hence we must have $x = y$. We have shown that

(7) for all $\psi \in \Lambda_{\mathscr{L}(\alpha)} \cap \Sigma_{\mathscr{L}(\alpha)}$, $\psi[(A, a)] = \psi[(B, b)]$.

It follows from Exercise 3I that the mapping $a_\upsilon \to b_\upsilon$ is an isomorphism on A onto the submodel of B generated by $\mathscr{R}b$, and hence A is isomorphically embeddable in B.

COROLLARY 7.2.9. The following three conditions are equivalent, for all $A, B \in \mathscr{M}$.

(i) A is isomorphically embeddable in some elementary extension B' of B.

(ii) For all $\varphi \in \Upsilon$, $\varphi[A]\, H \varphi[B]$.

(iii) For all $\varphi \in \Upsilon$, $\varphi[A] = \underset{\sim}{1}$ implies $\varphi[B] = \underset{\sim}{1}$.

PROOF. (i) implies (ii) by 7.2.5, By Exercise 1N we have $H[\underset{\sim}{1}] = \{\underset{\sim}{1}\}$, and hence (ii) implies (iii). Let B' be a special elementary extension of B of power $\geq \|A\|$, which exists by Theorems 6.2.9 and 7.2.6. By Theorem 7.2.8, (iii) implies that A is isomorphically embeddable in B', and so (iii) implies (i).

COROLLARY 7.2.10. If B is a special model of power α, then B is α-universal with respect to the theory

$$\Theta = \{h \in X^{\Sigma}: \breve{h}[\underset{\sim}{1}] \cap T \subset \breve{g}[\underset{\sim}{1}]\},$$

where $g = [B]$.

PROOF. This corollary is a restatement of Theorem 7.2.8.

A class $K \subset \mathscr{M}$ is an *existential class* if it is an intersection of finite unions of classes of the form

$$\{A \in \mathscr{M}: \varphi[A] \in H[Y]\}$$

where φ is an existential sentence and Y is closed in X. It is clear that every existential class is an elementary class, because by Lemma 1.4.1, $H[Y]$ is closed whenever Y is closed.

THEOREM 7.2.11. Let $K \in EC_{\Delta}$. Then K is an existential class if and only if $A \in K$ and $A \subset B$ imply $B \in K$.

PROOF. We first prove the so-called easy direction. Assume that K is an existential class. We wish to show that every extension of a member of K belongs to K. It is clearly sufficient to show this for the case in which K is a singular existential class, that is,

$$K = \{A \in \mathscr{M}: \varphi[A] \in H[Y]\}$$

for some existential sentence φ and closed set Y in X. Let $A \in K$ and $A \subset B$. By 7.2.5, $\varphi[A]H\varphi[B]$. Since H is transitive, $\varphi= [B] \in H[H[Y]] = H[Y]$. Hence $B \in K$.

Next assume that any extension of a member of K belongs to K. Let

$$L = \bigcap\{K' : K' \text{ is an existential class and } K \subset K'\}.$$

Clearly $K \subset L$ and L is an existential class. The theorem will be proved if we show that $L \subset K$. Suppose that $A \in L$. We first prove

(1) $$Th(A) \restriction T \in H = Th(K) \restriction T] .$$

Since $K \in EC_{\Delta}$, $Th(K)$ is closed in X^{Σ}, and hence $Th(K) \restriction T$ is closed in X^{T}. Then by Exercise 1P, the set $H[Th(K) \restriction T]$ is closed in X^{T}. Hence to prove (1) it is sufficient to show that

(2) for every finite set of existential sentences $\{\varphi_1, \ldots, \varphi_n\}$ and sequence of open sets $U_1, \ldots, U_n$ in X such that

$$\varphi_i[A] \in U_i \quad \text{for} \quad i = 1, \ldots, n,$$

we have

$$(U_1 \times \ldots \times U_n) \cap H[Th(K) \restriction \{\varphi_1, \ldots, \varphi_n\}] \neq 0.$$

Suppose (2) fails. Then there are $\varphi_1, \ldots, \varphi_n \in T$ and $U_1, \ldots, U_n \in' \mathcal{S}$ such that $\varphi_i[A] \in U_i$ for $i = 1, \ldots, n$, and

$$(U_1 \times \ldots \times U_n) \cap H[Th(K) \restriction \{\varphi_1, \ldots, \varphi_n\}] = 0.$$

In other words,

$$\breve{H}[U_1 \times \ldots \times U_n] \cap Th(K) \restriction \{\varphi_1, \ldots, \varphi_n\} = 0.$$

Since H is defined pointwise over X^n, this means

(3) $$(\breve{H}[U_1] \times \ldots \times \breve{H}[U_n]) \cap Th(K) \restriction \{\varphi_1, \ldots, \varphi_n\} = 0.$$

Let $Y_1 = X - \breve{H}[U_1], \ldots, Y_n = X - \breve{H}[U_n]$. Since $\breve{H}$ is open, each set Y_i is closed in X. Because H is transitive, $Y_i = H[Y_i]$ for each i. For each $i \leq n$, define

$$K_i = \{C \in \mathcal{M} : \varphi_i[C] \in Y_i\},$$

and let $K' = K_1 \cup \ldots \cup K_n$. Clearly K' is an existential class. By (3), we see that $K \subset K'$. Hence $A \in K'$. But this is impossible because for each i, $\varphi_i[A] \in U_i \subset X - Y_i$, and so $A \notin K_i$. Hence (2) holds and (1) is proved.

The condition (1) means that there is a $B \in K$ such that

$$\varphi[B] \, H \varphi[A] \quad \text{for each} \quad \varphi \in \Upsilon \ .$$

By Corollary 7.2.9, B can be isomorphically embedded in an elementary extension A' of A. Since $K \in EC_\Delta$, K is closed under isomorphisms. By hypothesis, any extension of B belongs to K, and therefore $A' \in K$. Again, K is elementarily closed, so $A \in K$. The theorem is proved.

The two-valued logic ℓ satisfies all of the assumptions 7.2.1 - 7.2.4. Hence, all of our theorems are true there. Furthermore, due to the connective $\vee$, we see that every existential class $K \subset M_\ell$ is simply an intersection of classes of the form

$$\{A \in \mathscr{M}_\ell : \ \varphi[A] = 1\}$$

for some existential sentence φ.

The logics $\mathscr{L}$ mentioned in Examples 2.5.2 and 2.5.3 satisfy all the assumptions 7.2.1 - 7.2.4. Again, all of our theorems are true in these cases.

EXERCISE 7A. Let $\Psi \subset \Sigma$. Suppose that $G \subset \mathscr{M} \times \mathscr{M}$ and G satisfies the following condition:

> if $\varphi[A] H \varphi[B]$ for every $\varphi \in \Psi$, then there exists A', $B' \in \mathscr{M}$ such that $Th(A) = Th(A')$, $Th(B) = Th(B')$ and $A' G B'$.

Show that if $K \in EC_\Delta$ and $G[K] \subset K$ then K is an intersection of finite unions of classes of the form

$$\{A \in \mathscr{M} : \ \varphi[A] \in H[Y]\}$$

where $\varphi \in \Psi$ and Y is closed in X.

For the following set of exercises we shall assume, in addition to 7.2.1 - 7.2.4, the following:

7.2.12 The dual logic $\mathscr{L}(\underset{\sim}{1}, \underset{\sim}{0}, \breve{H})$ (see Section 2.1) has a t-set $\mathscr{T}'$, an $\breve{H}$-set $\mathscr{H}'$, a k-set $\mathscr{K}' \subset \mathscr{C}_H$, and an e-set $\mathscr{E}' \subset \mathscr{Q}_H$.

EXERCISE 7B*. A formula $\varphi \in \Phi$ is said to be *universal* if

$$\varphi \in ((\mathscr{Q}_H \cup \mathscr{C}_H) \cap \mathscr{F})(\mathscr{C} \cap \mathscr{F})\Lambda .$$

State and prove results analogous to Theorems 7.2.8 and 7.2.11 for universal sentences and submodels.

EXERCISE 7C. A formula $\varphi \in \Phi$ is said to be *universal-existential* if

$$\varphi \in ((\mathscr{Q}_H \cup \mathscr{C}_H) \cap \mathscr{F})((\mathscr{Q}_{\breve{H}} \cup \mathscr{C}_H) \cap \mathscr{F})(\mathscr{C} \cap \mathscr{F})\Lambda .$$

Suppose $B \subset A \subset B'$ and $B \prec B'$; if φ is a universal-existential sentence, then $\varphi[A]\ H\varphi[B]$.

EXERCISE 7D. K is a *universal-existential class* if K is an intersection of finite unions of classes of the form

$$\{A \in \mathscr{M}: \ \varphi[A] \in H[Y]\}$$

where φ is a universal-existential sentence and Y is a closed set in X. Show that if K is a universal-existential class then K is closed under unions of chains of models in K. (Hint: Exercises 7C and 5J.2).

EXERCISE 7E*. Suppose that $\|B\| \leq \|A\| = \alpha$, A is special, and

$$\varphi[A]\ H\varphi[B] \text{ for every universal-existential sentence } \varphi.$$

Prove the following:

(i) There is a B' such that $B \prec B'$ and $B \subset A \subset B'$. Hint: First show that if $b \in S^{\alpha}$ enumerates B, then there exists $a \in R^{\alpha}$ such that

$$\varphi[(A, a)]\ H\varphi[(B, b)] \text{ for every existential sentence } \varphi \text{ in } \Sigma_{\mathscr{L}(\alpha)}.$$

(ii) If B is also a special model of power $\alpha \geq \|\mathcal{L}\|$, then in (i) we may take $B \cong B'$. (Hint: Use the downward Löwenheim-Skolem theorem.)

(iii) There is a B' such that $B \prec B'$ and B' is the union of a chain of models each of which is isomorphic to A.

EXERCISE 7F. Let $K \in EC_\Delta$. The following conditions are equivalent.

(i) K is a universal-existential class.

(ii) K is closed under unions of chains.

(iii) If $A \in K$, $B \subset A \subset B'$, and $B \prec B'$, then $B \in K$.

EXERCISE 7G*. What are the analogs of Exercise 7F when one considers sentences φ where

$$\varphi \in ((\mathcal{Q}_H \cup \mathcal{C}_H) \cap \mathcal{F})((\mathcal{Q}_{\check{H}} \cup \mathcal{C}_H) \cap \mathcal{F})((\mathcal{Q}_H \cup \mathcal{C}_H) \cap \mathcal{F}) \ldots (\mathcal{C} \cap \mathcal{F})\Lambda,$$

with n iterations? For a comprehensive treatment of this problem in the logic ℓ, see Keisler [1960].

7.3 Homomorphisms and positive classes

Throughout this section we shall assume the following:

7.3.1. $\langle \underset{\sim}{0}, \underset{\sim}{1} \rangle \in H$.

7.3.2. $\mathcal{L}$ has an H-set $\mathcal{H}$, a k-set $\mathcal{K} \subset \mathcal{C}_H$, and an e-set $\mathcal{E} \subset \mathcal{Q}_{H*}$.

7.3.3. The dual logic $\mathcal{L}' = \mathcal{L}(\underset{\sim}{1}, \underset{\sim}{0}, \check{H})$ (see §2.1 for definition) has an $\check{H}$-set $\mathcal{H}'$, a k-set $\mathcal{K}' \subset \mathcal{C}_H$, and an e-set $\mathcal{E}' \subset \mathcal{Q}_{H*}$.

Our assumptions for this section are different form those in §7.2 in that condition 7.2.1 has been dropped, conditions 7.2.2 and 7.2.3 are a part of 7.3.2, and a weaker version of 7.2.4 is also a part of 7.3.2. Of course, 7.3.1 has been added, and an entirely dual set of conditions has been added as 7.3.3. We shall apply throughout this section all of our earlier results either to the logic $\mathcal{L}$ or to the dual $\mathcal{L}'$. In particular, since conditions 7.1.1 - 7.1.3 hold for both $\mathcal{L}$ and $\mathcal{L}'$, all of our results in §7.1 hold for both $\mathcal{L}$ and $\mathcal{L}'$. We remind the reader that the dual of any result is achieved by interchanging the symbols $\underset{\sim}{0}$ and $\underset{\sim}{1}$, H and $\breve{H}$, $\mathscr{H}$ and $\mathscr{H}'$, $\mathscr{K}$ and $\mathscr{K}'$, and $\mathscr{E}$ and $\mathscr{E}'$.

The example 2.5.1 clearly gives rise to a pair ℓ and $\ell(\underset{\sim}{1}, \underset{\sim}{0}, \breve{H}_\ell)$ which satisfies 7.3.1 - 7.3.3. By some inessential changes in the sets $\mathscr{F} \cap \mathscr{C}$ and $\mathscr{F} \cap \mathscr{Q}$ the examples 2.5.2 and 2.5.3 also give rise to pairs of continuous logics $\mathcal{L}$ and $\mathcal{L}(\underset{\sim}{1}, \underset{\sim}{0}, \breve{H})$ which satisfy 7.3.1 - 7.3.3.

EXERCISE 7H. (i) Show that in examples 2.5.2 and 2.5.3 there are homeomorphisms F of X onto X such that xHy iff F(y)HF(x). Use this to prove the assertions made in the last paragraph.

(ii) Give an example of a logic $\mathcal{L}$ which satisfies 7.3.1 - 7.3.3 but there is no homeomorphism F of the sort described in (i).

A formula $\varphi \in \Phi$ is said to be *positive* if

$$\varphi \in ((\mathscr{Q}_{H*} \cup \mathscr{C}_H) \cap \mathscr{F})\wedge .$$

By the remarks at the beginning of §7.2, we see our notion of positive formulas coincides with the usual notion of positive formulas in the logic ℓ. Hence, for the continous logic $\mathcal{L}$ our notion may be regarded as a generalization of the usual notion in ℓ.

We let Π^+ be the set of all positive formulas in Σ^+, and let Π be the set of all positive sentences in Σ. Similarly, we let Π^+_μ and Π_μ denote the sets of all positive formulas in $\Sigma^+_{\mathcal{L}(\mu)}$ and $\Sigma_{\mathcal{L}(\mu)}$, respectively.

We reformulate Exercise 3F as follows:

7.3.4. If $\varphi \in \Pi^+$, AHB under the homomorphism h, and $r_0 \in R$, then $\varphi[A, r_0]\, H\varphi[B, h(r_0)]$. If $\varphi \in \Pi$ and AHB, then $\varphi[A]\, H\varphi[B]$.

THEOREM 7.3.5. Suppose that B is finite and for every positive sentence φ, $\varphi[B]\, H\varphi[A]$. Then $\|A\| \leq \|B\|$.

PROOF. Since $\underset{\sim}{0}$ is a continuous fixed point of $\breve{H}$, The hypothesis implies

(1) for every positive sentence φ, $\varphi[A] = \underset{\sim}{0}$ implies $\varphi[B] = \underset{\sim}{0}$.

Suppose $1 < n < \omega$ and $n \leq \|A\|$. Let $m = \frac{n(n-1)}{2}$. Since $\mathscr{L}'$ has a k-set $\mathscr{K}'$, we apply the dual of Lemma 2.4.7 to show that there exists $k \in \mathcal{C}_m \cap \mathscr{F}$ such that k is a composition of members of $\mathscr{K}'$ and

$$k(\underset{\sim}{0}, \ldots, \underset{\sim}{0}) = \underset{\sim}{0} \text{ and } \breve{k}[\underset{\sim}{0}] \subset (X - \{\underset{\sim}{1}\})^m.$$

Let φ_0 be the formula

$$\varphi_0 = k((v_1 \equiv v_2), (v_1 \equiv v_3), \ldots, (v_1 \equiv v_n), \ldots, (v_{n-1} \equiv v_n)).$$

Borrowing some terminology form the proof of Theorem 7.2.6, we see that φ_0 has the following properties:

(2) φ_0 is positive;

(3) for any $C \in \mathscr{M}$, $\varphi_0[C, r] = \underset{\sim}{0}$ for some $r \in T^\infty$ iff $n \leq \|C\|$;

(4) $\mathrm{val}(\varphi_0)$ is a finite set.

Now, using exactly the same argument as in the proof of Theorem 7.2.6, but applied to the logic $\mathscr{L}'$, we obtain a positive sentence φ_n which satisfies (2) and (3). Since $n \leq \|A\|$, it follows that $\varphi_n[A] = \underset{\sim}{0}$. So by (1), $\varphi_n[B] = \underset{\sim}{0}$ and hence $n \leq \|B\|$. This proves the theorem.

The next lemma needs no proof.

LEMMA 7.3.6. $\Pi^+ = \mathscr{E}(\mathscr{K}(\mathscr{H}(\Pi^+))) = \mathscr{E}'(\mathscr{K}'(\mathscr{H}'(\Pi^+)))$.

THEOREM 7.3.7. Suppose that A, B are special models, that either B is finite or $\|A\| = \|B\|$, and that $\varphi[A] \, H \varphi[B]$ for every positive sentence φ. Then AHB.

PROOF. The proof is an easy modification of the proof of Theorem 6.4 1. We first give the proof assuming that $\omega \leq \|A\| = \|B\|$. Then we shall indicate the proof when either A or B is finite.

Let $\omega \leq \alpha = \|A\| = \|B\|$ and let A_β, $\beta < \alpha$, and B_β, $\beta < \alpha$, be specializing chains for A, B, respectively. We choose enumerations $a \in R^\alpha$ and $b \in S^\alpha$ of R and S such that for all $\nu \in \alpha$,

$$a_\nu \in R_{|\nu|} \quad \text{and} \quad b_\nu \in S_{|\nu|}.$$

In what follows, we reserve the letter η for limit ordinals including 0. We shall construct sequences $d \in R^\alpha$ and $e \in S^\alpha$ such that for each $\nu < \alpha$, with $|\nu| = \gamma$, the following hold:

(1) $d_\nu \in R_\gamma$ and $e_\nu \in S_\gamma$;

(2) if $\varphi \in \Pi_{\nu+1}$ then $\varphi[(A, d\restriction\nu + 1)]\ H\varphi[(B, e\restriction\nu +)]$;

(3) if $\nu = \eta + 2n$, then $d_\nu = a_{\eta+n}$, and
if $\nu = \eta + 2n + 1$, then $e_\nu = b_{\eta+n}$.

We note first that by Exercise 3B applied to the logics $\mathscr{L}((\mathscr{C}_H \cup \mathscr{Q}_{H*}) \cap \mathscr{F})$ and $\mathscr{L}(\underset{\sim}{1}, \underset{\sim}{0}, (\mathscr{C}_H \cup \mathscr{Q}_{H*}) \cap \mathscr{F})$ condition (2) is equivalent to each of (2') and (2") below:

(2') if $\varphi \in \Pi_{\nu+1}$ and $\varphi[(A, d\restriction\nu + 1)] = \underset{\sim}{1}$, then
$\varphi[(B, e\restriction\nu + 1)] = \underset{\sim}{1}$;

(2") if $\varphi \in \Pi_{\nu+1}$ and $\varphi[(B, e\restriction\nu + 1)] = \underset{\sim}{0}$, then
$\varphi[(A, d\restriction\nu + 1)] = \underset{\sim}{0}$.

Suppose $\mu < \alpha$, $|\mu| = \beta$, and (1) - (3) hold for all $\nu < \mu$. Then by (2) we have

(4) $\varphi[(A, d\restriction\mu)]\ H\varphi[(B, e\restriction\mu)]$ for all $\varphi \in \Pi_\mu$, and by (1),

(5) $d\restriction\mu \in R^\mu_\beta$ and $e\restriction\mu \in S^\mu_\beta$.

Since $A_\beta \prec A$ and $B_\beta \prec B$, from (4) and (5) we get

(6) $\varphi[(A_\beta, d\restriction\mu)]\ H\varphi[(B_\beta, e\restriction\mu)]$ for all $\varphi \in \Pi_\mu$.

Suppose $\mu = \eta + 2n$. Define $d_\mu = a_{\eta+n}$. Condition (6) has an equivalent (6'), just like (2'), which gives us

(7) $\underset{\sim}{1} \in Th((B_\beta, e\restriction\mu))\restriction h^{\smile}[\underset{\sim}{1}]$ for the unique $h \in X^{\Pi_\mu}$ such that $h \in Th((A_\beta, d\restriction\mu))\restriction\Pi_\mu$.

Hence, by Lemma 7.3.6 and Theorem 7.1.6 applied to the logic $\mathcal{L}(\mu)$ and the set $\Psi = \Pi_\mu$, (7) yields

(8) $$\underset{\sim}{1} \in \overline{Th^+((B_\beta, e\restriction\mu))\restriction h^{\smile}[\underset{\sim}{1}]} \text{ for all } h \in Th^+((A_\beta, d\restriction\mu))\restriction\Pi^+_\mu.$$

Let $h_0 = [(A_\beta, d\restriction\mu), d_\mu] \restriction\Pi^+_\mu$. Then $h_0 \in Th^+((A_\beta, d\restriction\mu))\restriction\Pi^+_\mu$. Since B_β is β^+-saturated, $Th^+((B_\beta, e\restriction\mu))$ is closed. So (8) yields

(9) $$\underset{\sim}{1} \in Th^+((B_\beta, e\restriction\mu))\restriction h_0^{\smile}[\underset{\sim}{1}].$$

From (9) we may choose $e_\mu \in S_\beta$ such that

$$\text{if } \varphi \in \Pi^+_\mu \text{ and } \varphi[(A_\beta, d\restriction\mu), d_\mu] = \underset{\sim}{1}, \text{ then}$$
$$\varphi[(B_\beta, e\restriction\mu), e_\mu] = \underset{\sim}{1}.$$

Hence (2') and (2) hold for μ.

On the other hand, suppose that $\mu = \eta + 2n + 1$. Let $e_\mu = b_{\eta+n}$. We now argue in a similar manner but with respect to the logic $\mathcal{L}'$ to obtain an element $d_\mu \in R_\beta$ such that (2") and hence (2) hold for μ. Thus d and e are defined by transfinite induction. Clearly d and e enumerate R and S, respectively, and

$$\varphi[(A, d)] \, H\varphi[(B, e)] \text{ for all } \varphi \in \Pi_\alpha.$$

By Exercise 3I, AHB

Finally, suppose either A or B is finite. Let $\alpha = \|A\|$ and $\beta = \|B\|$. By Theorem 7.3.5, $\beta \leq \alpha$ and $\beta < \omega$. Let $b \in S^\beta$ be an enumeration of S. Regardless of whether α is finite or infinite, we find $a \in R^\beta$ such that

$$\text{if } \varphi \in \Pi_\beta \text{ and } \varphi[(A, a)] = \underset{\sim}{1}, \text{ then } \varphi[(B, b)] = \underset{\sim}{1}.$$

Now let $d \in R^\alpha$ be any enumeration of R. Since B is finite, B is

α-saturated. Hence, by the same technique as above, we find $e \in S^{\alpha}$ such that

$$\text{if } \varphi \in (\Pi_{\beta})_{\alpha} \text{ and } \varphi[(A, a), d] = 1, \text{ then } \varphi[(B, b), e] = 1.$$

This again implies AHB. The theorem is proved.

COROLLARY 7.3.8. The following three conditions are equivalent for all A, $B \in \mathscr{M}$.

(i) For some $A' \succ A$ and $B' \succ B$, we have A'HB'.

(ii) For all $\varphi \in \Pi$, $\varphi[A]\, H \varphi[B]$.

(iii) For all $\varphi \in \Pi$, $\varphi[A] = \underset{\sim}{1}$ implies $\varphi[B] = \underset{\sim}{1}$.

PROOF. The implication (i) to (ii) is trivial, and the equivalence of (ii) and (iii) follows from Exercise 3B. So let us assume (ii) and prove (i). Suppose that A is finite, then be Theorem 7.3.5, B is finite and $\|B\| \leq \|A\|$. Hence, by Theorem 7.3.7, AHB. If A is infinite, then regardless of whether B is finite or not there will exist special elementary extensions $A' \succ A$ and $B' \succ B$ such that the hypothesis 7.3.7 holds for A' and B'. Again, it follows from Theorem 7.3.7 that A'HB'. So (i) holds and the corollary is proved.

A class $K \subset \mathscr{M}$ is a *positive class* if it is an intersection of finite unions of classes of the form

$$\{A \in \mathscr{M}\colon \ \varphi[A] \in H[Y]\}$$

where φ is a positive sentence and Y is closed in X. Again since H[Y] is closed in X, every positive class is an elementary class.

THEOREM 7.3.9. Let $K \in EC_{\Delta}$. Then K is a positive class if and only if $A \in K$ and AHB imply $B \in K$.

PROOF. It is easy to verify that each class of the form

$$\{A \in \mathscr{M}\colon \ \varphi[A] \in H[Y]\}$$

where $\varphi \in \Pi$ and Y is closed in X is closed under taking homomorphic images. Hence each positive class K is closed under taking homomorphic images.

Next, assume that $A \in K$ and AHB imply $B \in K$. We can now apply Exercise 7A. Let the set Π replace the set Ψ and let the relation AHB replace the relation G in Exercise 7A. Then by Corollary 7.3.8 the hypothesis of Exercise 7A is satisfied. Hence, the conclusion of Exercise 7A must hold, and this proves that K is a positive class. We add here for those readers who have not proved Exercise 7A that a proof for this part of the theorem may be found by copying word for word the corresponding proof in Theorem 7.2.11, with T everywhere replaced by Π and "existential" by "positive". Exercise 7A is the abstraction of this process. The theorem is proved.

As we have remarked earlier in this section, the logics ℓ and $\ell(\underset{\sim}{1}, \underset{\sim}{0}, \breve{H}_\ell)$ satisfy all of the assumptions 7.3.1 - 7.3.3. Hence all of our results hold there. Again, due to the connective $\otimes$, we see that every positive class $K \subset \mathcal{M}_\ell$ is simply an intersection of classes of the form

$$\{A \in \mathcal{M}_\ell : \varphi[A] = \underset{\sim}{1}\}$$

where φ is a positive sentence.

A large number of exercises may now be formulated in the spirit of the exercises found at the end of §7.2. We present one as a sample.

EXERCISE 7I. A formula φ is said to be *positive existential* iff

$$\varphi \in ((\mathcal{Q}_{\breve{H}} \cup \mathcal{C}_H) \cap \mathcal{F})\Lambda.$$

A class $K \subset \mathcal{M}$ is a *positive existential* class if K is an intersection of finite unions of classes of the form

$$\{A \in \mathcal{M}: \varphi[A] \in H[Y]\}$$

where φ is a positive existential sentence and Y is closed in X. Assume for the logics $\mathcal{L}$ and $\mathcal{L}'$ the conditions 7.2.1–7.2.4, 7.2.12, and 7.3.1. Let $K \in EC_\Delta$. Then the following three conditions are equivalent:

(i) K is a positive existential class.

(ii) If $A \in K$, AHB, and $B \subset C$, then $C \in K$.

(iii) If $A \in K$, $A \subset B$, and BHC, then $C \in K$.

7.4. Reduced products and conditional classes

The content of this section is somewhat more special than the results of the preceding two sections. We shall present, in as general a setting as possible, the result of Keisler [1965] on reduced products and conditional classes for the logic ℓ. In order to define a meaningful reduced product, we shall have to put further restrictions on the binary relation H which orders the space X and on the continuous logic $\mathcal{L}$. We shall require throughout this section that:

7.4.1. H is a simple order relation over X.

7.4.2. $\mathcal{L}$ has an H-set $\mathcal{H}$.

7.4.3. There exists a homeomorphism $t \in \mathcal{F}$ of X onto X such that xHy iff t(y)Ht(x).

We see that examples 2.5.1 and 2.5.2 still satisfy all of our assumptions, however example 2.5.3 fails to satisfy 7.4.1.

Actually, for our main result the set $\mathcal{F}$ will have to contain some other functions than those listed in 7.4.2 and 7,4.3. It turns out that 7.4.1 and 7.4.2 already imply that these functions will be found in $\mathcal{C}$ and $\mathcal{Q}$. So, we shall first explore some consequences of 7.4.1 and 7.4.2, and then we shall make some more assumptions about $\mathcal{L}$ (see 7.4.9 and 7.4.14).

First we observe that

7.4.4. $\underset{\sim}{0}H\underset{\sim}{1}$; for all $x \in X$, $\underset{\sim}{0}Hx$ and $xH\underset{\sim}{1}$; $t(\underset{\sim}{0}) = \underset{\sim}{1}$ and $t(\underset{\sim}{1}) = \underset{\sim}{0}$; the dual logic $\mathcal{L}' = \mathcal{L}(\underset{\sim}{1}, \underset{\sim}{0}, \breve{H})$ has an $\breve{H}$-set $\mathcal{H}'$.

Next, by the definition of an H-set, the following hold.

7.4.5. If $\langle x, y\rangle \in X^2 - H$, then there exist disjoint open sets U and V such that

$$H[U] = U, \quad \breve{H}[V] = V, \quad x \in U, \quad \text{and} \quad y \in V.$$

From 7.4.5 and its dual we see that

7.4.6. $X^2 - H$ and $X^2 - \breve{H}$ are open in X^2, and hence H and $\breve{H}$ are closed in X^2.

For the remainder of this section we shall write

$$x \leq y \quad \text{for} \quad xHy$$

and

$$x < y \quad \text{for} \quad xHy \quad \text{and} \quad x \neq y.$$

We do this for reasons of clarity and economy. We can always tell by context whether the symbols $\leq$ and $<$ refer to the ordering on the ordinals or to H. With respect to the newly introduced notion we define the usual closed, open, and half-open intervals as follows:

$$[x, y] = \{z \in X: \; x \leq z \leq y\},$$
$$(x, y) = \{z \in X: \; x < z < y\},$$
$$[x, y) = \{z \in X: \; x \leq z < y\},$$
$$(x, y] = \{z \in X: \; x < z \leq y\}.$$

From Lemma 1.4.1 we see that

7.4.7. Any closed interval is closed in X and any open interval is open in X. Furthermore, the closed interval $[\underset{\sim}{0}, \underset{\sim}{1}]$ and the half-open intervals $[\underset{\sim}{0}, x)$ and $(x, \underset{\sim}{1}]$ are open in X.

We define the functions max and min on X^2 with respect to the ordering H in the usual way.

LEMMA 7.4.8. max, min $\in \mathcal{C}_2 \cap \mathcal{C}_H$.

PROOF. Let us show that $\min \in \mathcal{C}_2$. Let V be open in X. Then

$$\min^{\smile}[V] = (V \times X \cap X^2 - \breve{H}) \cup (X \times V \cap X^2 - H) \cup V \times V,$$

which is open in X^2.

We henceforth assume that

7.4.9. $\max, \min \in \mathcal{F}$.

It is immediate that now $\mathcal{L}^{\circ}$ has a k-set $\{\min\}$ and $\mathcal{L}^{\circ\prime}$ has a k-set $\{\max\}$. Since both min and max are associative, we shall write

$$\min(x_1, \ldots, x_n) \text{ and } \max(x_1, \ldots, x_n)$$

for $\min\{x_1, \ldots, x_n\}$ and $\max\{x_1, \ldots, x_n\}$, respectively.

LEMMA 7.4.10. Let Y be a non-empty closed set in X. Then there exists a $y \in Y$ such that $H[y] = H[Y]$.

PROOF. Suppose that, to the contrary, for each $y \in Y$ there exists a $z_y \in Y$ such that $z_y < y$. By the dual of 7.4.5, there exists an open set V_y such that

$$H[V_y] = V_y, \quad y \in V_y, \quad \text{and} \quad z_y \notin V_y.$$

The collection $\{V_y: y \in Y\}$ is an open covering of Y, hence by compactness there are $y_1, \ldots, y_n \in Y$ such that

$$Y \subset V_{y_1} \cup \ldots \cup V_{y_n}.$$

Let $z = \min(z_{y_1}, \ldots, z_{y_n})$. Clearly $z \in Y$ but

$$z \notin V_{y_1} \cup \ldots \cup V_{y_n}.$$

This is a contradiction. The lemma is proved.

By Lemma 7.4.10 and its dual, we define two functions, inf and sup, with domain X^* as follows: For $Y \in X^*$,

$\inf Y$ = the unique $y \in \overline{Y}$ such that $H[y] = H[\overline{Y}]$

$\sup Y$ = the unique $y \in \overline{Y}$ such that $\breve{H}[y] = \breve{H}[\overline{Y}]$.

LEMMA 7.4.11. Let $Y \in X^*$. Then $\inf Y$ and $\sup Y$ are, respectively, the g.l.b. and l.u.b. of the set Y.

PROOF. We show that $\inf Y$ is the g.l.b. of Y. Since $Y \subset \overline{Y}$ and $H[\inf Y] = H[\overline{Y}]$, we have that $\inf Y$ is a lower bound for Y. Suppose y is a lower bound for Y. Hence $Y \subset H[y]$. Since $H[y]$ is closed, $\overline{Y} \subset H[y]$. So $H[\overline{Y}] \subset H[y]$, $H[\inf Y] \subset H[y]$, and $y \leq \inf Y$.

LEMMA 7.4.12. Suppose that $x \in V \in \mathcal{S}$. Then there exists an open set W which is an interval of the form $[\underset{\sim}{0}, \underset{\sim}{1}]$, or $[\underset{\sim}{0}, y)$, or $(y, \underset{\sim}{1}]$, or (y, z) such that $x \in W$ and $W \subset V$. That is, the above intervals form a base for the topology $\mathcal{S}$.

PROOF. Consider the closed sets

$$Y_1 = \breve{H}[x] \cap (X - V),$$
$$Y_2 = H[x] \cap (X - V).$$

If $Y_1 = Y_2 = 0$, then let $W = [0, 1]$. Suppose that $Y_1 \neq 0$ and $Y_2 \neq 0$. Let $y_1 = \sup Y_1$ and $y_2 = \inf Y_2$. Since Y_1, Y_2 are closed, $y_1 \in Y_1$ and $y_2 \in Y_2$. By Lemma 7.4.11, we have

$$y_1 < x < y_2.$$

Consider the open interval $W = (y_1, y_2)$. Clearly $x \in W$ and $W \subset V$. In case either $Y_1 = 0$ or $Y_2 = 0$, a similar argument shows that W may be taken to be a half-open interval of the form $[\underset{\sim}{0}, y)$ or $(y, \underset{\sim}{1}]$.

LEMMA 7.4.13. $\sup, \inf \in \mathcal{C}$.

PROOF. We shall only show that $\sup \in \mathcal{C}$. Let V be an open set. By Lemma 7.4.12, we need only show that $\sup^{\smile}[V]$ is open in X^* when V is an interval which is open in X. Without loss of generality let us assume that $V = (x, y)$. We assert that for $Y \in X^*$,

$$\sup Y \in V \text{ iff } \overline{Y} \subset \breve{H}[V] \text{ and } Y \cap V \neq 0.$$

Since $\check{H}[V]$ is open, this says that $\sup^{\smile}[V]$ is open in X^*. The lemma is proved.

We henceforth assume that

7.4.14. $\inf, \sup \in \mathscr{F}$.

It is evident that now $\mathscr{L}$ has an e-set $\{\sup\}$ and $\mathscr{L}'$ has an e-set $\{\inf\}$.

We now turn to the problem of defining a reduced product. Let D be a filter over I, a non-empty index set, and let $g \in X^I$. Let

$$D(g) = \{E\text{-lim } g\colon\ D \subset E \text{ and } E \text{ is an ultrafilter over } I\}.$$

LEMMA 7.4.15. $D(g)$ is closed in X.

PROOF. We show that $X - D(g)$ is open. Let $x \in X - D(g)$. We claim that

(1) there exists an open set U such that $x \in U$ and $\{i\colon\ g(i) \notin U\} \in D$.

Suppose, to the contrary,

(2) for every open set U such that $x \in U$, we have

$$\{i\colon\ g(i) \notin U\} \notin D.$$

Let

$$E' = \{\{i\colon\ g(i) \in U\}\colon\ x \in U \in \mathscr{S}\}.$$

Clearly by (2) E' has the finite intersection property. Furthermore, using (2) once again,

$E' \cup D$ has the finite intersection property.

Now, let E be an ultrafilter over I such that $E' \cup D \subset E$. Then $x = E\text{-lim } g$, a contradiction to $x \notin D(g)$. Hence (1) holds. Clearly no point in U can belong to $D(g)$. So $U \cap D(g) = 0$ and the lemma is proved.

We define the *lower D-limit of* g, written D-inf g, as follows:

$$\text{D-inf } g = \inf D(g) \quad .$$

Notice that if D is an ultrafilter then D-inf g = D-lim g. Also, by Lemma 7.4.15, for every $g \in X^I$, there exists an ultrafilter $E \supset D$ such that D-inf g = E-lim g.

The next lemma proves to be very useful.

LEMMA 7.4.16. D-inf g = $\sup\{\inf\{g(i): i \in J\}: J \in D\}$.

PROOF. Let

$$x = \text{D-inf } g \quad \text{and} \quad y = \sup\{ \inf\{g(i): i \in J\}: J \in D\}.$$

Suppose first that $x < y$. Then there exists a $J \in D$ such that

$$x < \inf\{g(i): i \in J\}.$$

By Lemma 7.4.5 there exists an open set U such that

$$\breve{H}[U] = U, \quad \text{and} \quad \inf\{g(i): i \in J\} \notin U.$$

This implies that

$$\{i: g(i) \notin U\} \supset J \in D.$$

So $x \notin D(g)$, a contradiction. Next suppose that $y < x$. Since $x = \inf D(g)$, we have $y \notin D(g)$. Hence by condition (1) in the proof of Lemma 7.4.15, there exists an open set U and a set $J_1 \subset I$ such that

$$y \in U \quad \text{and} \quad J_1 = \{i: g(i) \notin U\} \in D.$$

By Lemma 7.4.12, we may assume that U is an interval. Moreover, since $y \neq \underset{\sim}{1}$, we may assume that U is open on the right. That is either

(1) $$U = [0, z_2)$$

or

(2) $$U = (z_1, z_2).$$

In case (1),

$$y < z_2 \leq \inf\{g(i): i \in J_1\},$$

which is a contradiction to the definition of y. In case (2), since $z_1 < y$, there exists $J_2 \in D$ such that

$$z_1 < \inf\{g(i): i \in J_2\}.$$

Let $i \in J_1 \cap J_2$. Then $g(i) \notin U$ and $z_1 < g(i)$. This implies $z_2 \leq g(i)$. Hence again we have the contradiction

$$y < z_2 \leq \inf\{g(i): i \in J_1 \cap J_2\}.$$

So we must have $x = y$ and the lemma is proved.

The following exercise shows that our definition of reduced product will coincide with the usual definition in the logic ℓ.

EXERCISE 7J. Suppose that $\sup(\mathscr{R}g - \{\underset{\sim}{1}\}) \neq \underset{\sim}{1}$. Then D-inf $g = \underset{\sim}{1}$ iff $\{i: g(i) = \underset{\sim}{1}\} \in D$.

Let $F = \lambda i A_i$ be a function on I into $\mathscr{M}$. Let us recall the following definitions from §1.2: let g, h range over $\Pi_{i \in I} R_i$,

$$g \sim h \text{ iff } \{i \in I: g(i) = h(i)\} \in D;$$

$$g^{\sim} = \{h: h \sim g\};$$

$$\text{D-prod } \lambda i R_i = \{g^{\sim}: g \in \Pi_{i \in I} R_i\}.$$

The *reduced product of the models* A_i, $i \in I$, also written D-prod $\lambda i A_i$, is the model $A = (R, \mathscr{A})$ in $\mathscr{M}$ such that:

7.4.17. $R = \text{D-prod } \lambda i R_i$;

7.4.18. $\mathscr{A}(c_\zeta) = (\lambda i \mathscr{A}_i(c_\zeta))^{\sim}$ for each $\zeta < \kappa$;

7.4.19. for each $\eta < \pi$ and $g_1^{\sim}, \ldots, g_{\tau(\eta)}^{\sim} \in R$,

$$\mathscr{A}(P_\eta)(g_1^{\sim}, \ldots, g_{\tau(\eta)}^{\sim}) = \text{D-inf } \lambda i \mathscr{A}_i(P_\eta)(g_1(i), \ldots, g_{\tau(\eta)}(i)).$$

It is easily seen that each $\mathscr{A}(c_\zeta) \in R$. Moreover, since each $\tau(\eta)$ is finite, we can verify easily, using the characterization provided by Lemma 7.4.16, that each $\mathscr{A}(P_\eta)$ is a well defined function, independent of the representation of the equivalence classes. By Exercise 7J, we see that

$$(v_0 \equiv v_1)[A, r] = \text{D-inf } g$$

where $g \in \{\underset{\sim}{0}, \underset{\sim}{1}\}^I$, $g(i) = \underset{\sim}{0}$ if $r_0(i) \neq r_1(i)$, and $g(i) = \underset{\sim}{1}$ if $r_0(i) = r_1(i)$. Thus the treatment of all formulas $\varphi \in \Lambda$ is uniform.

If $B = A_i$ for all $i \in I$, then we write D-prod B for D-prod $\lambda i A_i$ and call it the *reduced power of* B. If D is an ultrafilter over I, then it is clear that the two notions of reduced product and ultraproduct coincide. Furthermore, in case of the logic ℓ and the models in $\mathscr{M}_\ell$, our notion of reduced product specializes, again by Exercise 7J, down to the ordinary notion.

The Leibniz law for identity forces us to take 7.4.19 as the definition for $\mathscr{A}(P_\eta)(\tilde{g}_1, \ldots, \tilde{g}_{\tau(\eta)})$. Although we know by duality that (subject to proper definition)

$$\text{D-sup } \lambda i \mathscr{A}_i(P_\eta)(g_1(i), \ldots, g_{\tau(\eta)}(i))$$

also exists, we do not know whether it will lead to other interesting algebraic constructions. We do not know, for instance, how much of our construction of reduced product can be duplicated in case the ordering H on X is merely assumed to be a partial order relation. In view of the fact that $D(g) \in X^*$ and $D(g)$ is closed in X (Lemma 7.4.15), our construction may be a very special case of obtaining a model A with values in the compact Hausdorff space X^* (restricted to closed sets in X) from the models A_i with values in X.

We shall not develop here in detail the basic properties of reduced products, as we did for ultraproducts in Chapter 5. Such a detailed development in the case of the logic ℓ can be found in Frayne-Morel-Scott [1962]. We see no difficulty in duplicating most of the results in Frayne-Morel-Scott [1962], if we wish. However our main concern in this section is to characterize conditional classes in terms of reduced products For this purpose we need the following definitions.

A formula $\varphi \in \Phi$ is *primitively conditional*, in symbols $\varphi \in \Gamma'$, if there are atomic formulas $\varphi_1, \ldots, \varphi_{n+1} \in \Lambda$ and functions $t_1, \ldots, t_{n+1} \in \mathscr{H}$ such that either

$$\varphi = \max(t_1 t\varphi_1, \ldots, t_n t\varphi_n)$$

or

$$\varphi = \max(t_1 t\varphi_1, \ldots, t_n t\varphi_n, t_{n+1}\varphi_{n+1}).$$

φ is a *conditional formula* if

$$\varphi \in \{\inf, \sup, \min\}\Gamma'.$$

We let Γ^+ be the set of all conditional formulas in Σ^+, and let Γ be the set of all conditional sentences. Similarly, we let Γ_μ^+ and Γ_μ denote the sets of all conditional formulas in $\Sigma_{\mathscr{L}(\mu)}^+$ and $\Sigma_{\mathscr{L}(\mu)}$, respectively.

We remind the reader that in the logic ℓ,

$$t = \neg, \quad \sup = \exists, \quad \inf = \forall, \quad \text{and} \quad \mathscr{H} = \{\mathrm{id}\}.$$

Hence our notion of conditional formulas and sentences specializes to the usual notion of conditional (or Horn) formulas and sentences.

The next lemma is a part of the first step of the induction required to prove Exercise 7K.

LEMMA 7.4.20. Let $g_1, \ldots, g_{n+1} \in X^I$, $t_1, \ldots, t_{n+1} \in \mathscr{H}$, and $y_j = D\text{-}\inf g_j$ for $1 \leq j \leq n+1$. Suppose that

$$J_0 = \{i\colon \max(t_1 tg_1(i), \ldots, t_n tg_n(i), t_{n+1}g_{n+1}(i)) = \underset{\sim}{1}\} \in D.$$

Then

$$\max(t_1 ty_1, \ldots, t_{n+1}y_{n+1}) = \underset{\sim}{1}.$$

PROOF. Suppose that

(1) $$t_j t y_j < \underset{\sim}{1} \quad \text{for} \quad 1 \leq j \leq n.$$

We shall prove that $t_{n+1} y_{n+1} = \underset{\sim}{1}$. Let

(2) $$z_j = \inf t_j^{\vee}[\underset{\sim}{1}] \quad \text{for} \quad 1 \leq j \leq n + 1.$$

From (1) and (2), we see that

(3) $$t y_j < z_j \quad \text{for} \quad 1 \leq j \leq n.$$

Hence, applying $t^{\vee}$ to (3), we have

(4) $$t^{\vee}(z_j) < y_j \quad \text{for} \quad 1 \leq j \leq n.$$

By the definition of y_j and Lemma 7.4. 16, (4) implies

(5) for $1 \leq j \leq n$, there exists $J_j \in D$ such that
$$t^{\vee}(z_j) < \inf\{g_j(i): \; i \in J_j\}.$$

Let $J = J_0 \cap J_1 \cap \dots \cap J_n$. Certainly $J \in D$. If $i \in J$, then we have by (5) and the hypothesis,

(6) $$\max(t_1 t g_1(i), \; \dots \; , t_{n+1} g_{n+1}(i)) = \underset{\sim}{1}$$

and

(7) $$t^{\vee}(z_j) < g_j(i) \quad \text{for} \quad 1 \leq j \leq n.$$

Hence, applying t to (7),

$$t g_j(i) < z_j \quad \text{for} \quad 1 \leq j \leq n,$$

and by (2),

$$t_j t g_j(i) < \underset{\sim}{1} \quad \text{for} \quad 1 \leq j \leq n.$$

By (6), this implies $t_{n+1} g_{n+1}(i) = \underset{\sim}{1}$. Hence

$$\{i: \; t_{n+1} g_{n+1}(i) = \underset{\sim}{1}\} \in D,$$

and by (2),

$$\{i: \; z_{n+1} \leq g_{n+1}(i)\} \in D.$$

Using Lemma 7.4.16, this means

$$z_{n+1} \leq D\text{-inf } g_{n+1} = y_{n+1}.$$

So, $t_{n+1} y_{n+1} = \underset{\sim}{1}$, as was to be proved.

EXERCISE 7K. Suppose that $\varphi \in \Gamma$ and for each $i \in I$, $\varphi[A_i] = \underset{\sim}{1}$. Then

$$\varphi[D\text{-prod } \lambda i \; A_i] = \underset{\sim}{1}.$$

A class $K \subset \mathscr{M}$ is a *conditional class* if K is an intersection of classes of the form

$$\{A \in \mathscr{M}: \; \varphi[A] = \underset{\sim}{1}\}$$

where φ is a conditional sentence. Notice that we are not taking finite unions of classes of this form. Also, instead of an arbitrary closed set Y, we take the set $\{\underset{\sim}{1}\}$.

Exercise 7K already tells us that each conditional class K is closed under reduced products of members of K. The next few lemmas will lead to the converse of this result.

Recall from §1.1 that

$$S^{\alpha}(I) = \{J: \; J \quad I \text{ and } |I - J| < \alpha\}.$$

Temporarily we let $R' = \Pi_{i \in I} R_i$. For $r \in (R')^{\infty}$ and $i \in I$, r^i shall denote that function in $(R_i)^{\infty}$ such that

$$r^i(n) = r_n(i) \quad \text{for all } n < \omega .$$

We shall adopt this notation also for general sequences $d \in R'^{\beta}$.

LEMMA 7.4.21. Suppose that $\|\mathscr{L}\| \leq \alpha$, $\alpha^+ = 2^{\alpha}$, $|I| = \alpha$, $\|A_i\| \leq \alpha^+$ for each $i \in I$, and that

B is either finite or special and of power α^+. Furthermore, for every $\varphi \in \Gamma$,

$$\text{if } \{i:\ \varphi[A_i] = \underset{\sim}{1}\} \in S^\alpha(I), \text{ then } \varphi[B] = \underset{\sim}{1}.$$

Then there exists a mapping h of R' onto S such that for every conditional formula φ and every $r \in (R')^\infty$,

$$\text{if } \{i:\ \varphi[A_i, r^i] = \underset{\sim}{1}\} \in S^\alpha(I), \text{ then } \varphi[B, h \circ r] = \underset{\sim}{1}.$$

PROOF. Since $2^\alpha = \alpha^+$, we have $|R'| \leq \alpha^+$. Let $a \in (R')^{\alpha+}$ and $b \in S^{\alpha+}$ be enumerations of R' and S, respectively. Our proof will establish that, under the hypothesis of the theorem, $|R'| \geq \|B\|$, regardless of whether B is finite or of power α^+, but provided B is α^+-saturated. We construct two sequences $d \in R^{\alpha^+}$, $e \in S^{\alpha^+}$ such that for each $\nu < \alpha^+$,

(1) whenever $\varphi \in \Gamma_{\nu+1}$ and

$$\{i:\ \varphi[(A_i, d^i{\restriction}\nu + 1)] = \underset{\sim}{1}\} \in S^\alpha(I), \text{ then } \varphi[(B, e{\restriction}\nu + 1)] = \underset{\sim}{1};$$

(2) if η is a limit ordinal, including 0, then

if $\nu = \eta + 2n$, then $d_\nu = a_{\eta+n}$,

if $\nu = \eta + 2n + 1$, then $e_\nu = b_{\eta+n}$.

Let $\mu < \alpha^+$, and suppose (1) and (2) hold for all $\nu < \mu$. Then clearly we have,

(3) whenever $\varphi \in \Gamma_\mu$ and

$$\{i:\ \varphi[(A_i, d^i{\restriction}\mu)] = \underset{\sim}{1}\} \in S^\alpha(I), \text{ then } \varphi[(B, e{\restriction}\mu)] = \underset{\sim}{1}.$$

Suppose $\mu = \eta + 2n$. We define $d_\mu = a_{\eta+n}$. Let

(4) $\Delta = \{\varphi \in \Pi^+_\mu:\ \{i:\ \varphi[(A_i, d^i{\restriction}\mu),\ d_\mu(i)] = \underset{\sim}{1}\} \in S^\alpha(I)\}$.

We show that the constant function $\underset{\sim}{1} \in X^\Delta$ is a member of

$$\overline{Th^+((B, e{\restriction}\mu)){\restriction}\Delta}.$$

For this purpose, let $\varphi_1, \ldots, \varphi_m \in \Delta$ and $\underset{\sim}{1} \in V \in \mathscr{S}$. We show that

(5) there exists $s_0 \in S$ such that

$$\varphi_j[(B, e\restriction\mu), s_0] \in V \text{ for } 1 \leq j \leq m.$$

Let $\varphi = \min(\varphi_1, \ldots, \varphi_m)$. We have $\varphi \in \Gamma_\mu^+$ and $(\sup v_0)\, \varphi \in \Gamma_\mu$. By (4),

$$\{i: \varphi[(A_i, d^i\restriction\mu), d_\mu(i)] = \underset{\sim}{1}\} \in S^\alpha(I),$$

and hence

$$\{i: (\sup v_0)\, \varphi[(A_i, d^i\restriction\mu)] = \underset{\sim}{1}\} \in S^\alpha(I).$$

So, by (3), $(\sup v_0)\, \varphi[(B, e\restriction\mu)] = \underset{\sim}{1}$, and this means

(6) $$\sup\{\varphi[(B, e\restriction\mu), s]: s \in S\} = \underset{\sim}{1}.$$

(6) clearly implies that there exists an $s_0 \in S$ such that

$$\varphi[(B, e\restriction\mu), s_0] \in V.$$

This immediately gives (5). So

$$\underset{\sim}{1} \in \overline{Th^+(B, e\restriction\mu)\restriction\Delta}.$$

Since B is α^+-saturated, $Th^+(B, e\restriction\mu)$ is closed, so

$$\underset{\sim}{1} \in Th^+(B, e\restriction\mu)\restriction\Delta.$$

We now pick $e_\mu \in S$ so that

$$\varphi[(B, e\restriction\mu), e_\mu] = \underset{\sim}{1} \text{ for all } \varphi \in \Delta.$$

For this choice of e_μ, clearly (1) holds for μ.

Next suppose $\mu = \eta + 2n + 1$. We define $e_\mu = b_{\eta+n}$. Let

$$\Delta = \{\varphi \in \Gamma_\mu^+: \varphi[(B, e\restriction\mu), e_\mu] \neq \underset{\sim}{1}\}.$$

Hence

$$(\inf v_0)\, \varphi[(B, e\restriction\mu] \neq \underset{\sim}{1} \text{ for every } \varphi \in \Delta.$$

Since $(\inf v_0)\,\varphi \in \Gamma_\mu$, this implies by (3),

$$\{i: \ (\inf v_0)\,\varphi[(A_i, d^i \restriction \mu)] = 1\} \notin S^\alpha(I).$$

For each $\varphi \in \Delta$, let

$$I_\varphi = \{i: \ (\inf v_0)\,\varphi[(A_i, d^i \restriction \mu)] \neq 1\}.$$

By the above we see that

$$|I_\varphi| = \alpha \quad \text{for each} \quad \varphi \in \Delta.$$

Since $|\Delta| \leq \alpha$, we may apply Lemma 5.4.5 and obtain a family of subsets of I,

$$\{J_\varphi: \ \varphi \in \Delta\},$$

such that

$$J_\varphi \subset I_\varphi \quad \text{and} \quad |J_\varphi| = \alpha \quad \text{for each} \quad \varphi \in \Delta,$$

and

$$J_\varphi \cap J_\psi = 0 \quad \text{if} \quad \varphi \neq \psi.$$

Let $\varphi \in \Delta$ and $i \in J_\varphi$. Then $i \in I_\varphi$ and this means

$$(\inf v_0)\,\varphi[(A_i, d^i \restriction \mu)] \neq 1.$$

Hence, there exists $r_i \in R_i$ such that

$$\varphi[(A_i, d^i \restriction \mu), r_i] \neq 1.$$

Define d_μ as follows:

$$d_\mu(i) = r_i \quad \text{if} \quad \varphi \in \Delta \quad \text{and} \quad i \in J_\varphi,$$
$$d_\mu(i) \quad \text{is arbitrary otherwise.}$$

Notice that since the J_φ's are disjoint, d_μ is well-defined. Now, for this particular d_μ, (1) Holds with μ in place of ν. Hence the induction is complete. The mapping h is given by $h(d_\mu) = e_\mu$ for each $\mu < \alpha^+$. It can easily be verified that h is well-defined and that h

has domain R' and range S.

LEMMA 7.4.22. Assume the hypotheses of the previous lemma. Then there exists a filter D over I such that

$$D\text{-prod}\ \lambda i A_i \cong B.$$

PROOF. Let us assume that a function h satisfying the conclusion of Lemma 7.4.21 has been constructed. For each $\varphi \in \Lambda$, $r \in (R')^{\infty}$, and $y \in X$ such that

$$y < \varphi[B, h \circ r],$$

we define the set

$$I(\varphi, r, y) = \{i: \varphi[A_i, r^i] > y\}.$$

Let

$$E = \{I(\varphi, r, y): \text{ where } \varphi, r, y \text{ satisfy the above}\}.$$

We prove that

(1) E has the finite intersection property.

Let $J_1 = I(\varphi_1, r_1, y_1), \ldots, J_n = I(\varphi_n, r_n, y_n) \in E$. We may assume, by the elementary properties of substitution, that the φ_j's have been picked in such a manner thet $r = r_1 = r_2 = \ldots = r_n$. Since

$$y_j < \varphi_j[B, h \circ r], \quad \text{for } 1 \leq j \leq n$$

we have

$$t\varphi_j[B, h \circ r] < ty_j \quad \text{for } 1 \leq j \leq n.$$

For $1 \leq j \leq n$, let $t_j \in \mathscr{H}$ be such that

(2) $$t_j t\varphi_j[b, h \circ r] \neq \underset{\sim}{1} \quad \text{and} \quad t_j t y_j = \underset{\sim}{1}.$$

Consider the conditional formula

$$\varphi = \max(t_1 t\varphi_1, \ \ldots \ , t_n t\varphi_n).$$

If $i \notin J_j$, then $\varphi_j[A_i, r^i] \leq y_j$. Hence $ty_j \leq t\varphi_j[A_i, r^i]$ and so $t_j t\varphi_j[A_i, r^i] = \underset{\sim}{1}$. Thus,

$$\text{if } \ i \in \overline{J}_1 \cup \ldots \cup \overline{J}_n, \ \text{ then } \ \varphi[A_i, r^i] = \underset{\sim}{1}.$$

If $|J_1 \cap \ldots \cap J_n| < \alpha$, then $\overline{J}_1 \cup \ldots \cup \overline{J}_n \in S^{\alpha}(I)$, and by the definition of h, $\varphi[B, h \circ r] = \underset{\sim}{1}$. This means that

(3) $$t_j t\varphi_j[B, h \circ r] = \underset{\sim}{1} \ \text{ for some } \ j, \ 1 \leq j \leq n.$$

(3) is impossible in view of (2). So $|J_1 \cap \ldots \cap J_n| = \alpha$ and (1) is proved.

Let D be the filter over I generated by E. Let $A = D\text{-prod } \lambda i A_i$. We prove that $A \cong B$. In what follows we shall consider the identity, $\equiv$, to be among the symbols P_η, $\eta < \pi$. Let $\eta < \pi$, $n = \tau(\eta)$, $g_1, \ \ldots \ , g_n \in R'$,

and let

$$x = D\text{-inf } \lambda i \ \mathscr{A}_i(P_\eta)(g_1(i), \ \ldots \ , g_n(i)),$$
$$y = \mathscr{B}(P_\eta)(hg_1, \ \ldots \ , hg_n).$$

For our purpose it is sufficient to establish that $x = y$. If $x < y$, then

(4) $\inf\{\mathscr{A}_i(P_\eta)(g_1(i), \ \ldots \ , g_n(i)): \ i \in J\} \leq x$ for all $J \in D$.

Suppose first there exists a $z \in X$ such that $x < z < y$. Let

$$J_0 = I(P_\eta(v_1, \ \ldots \ , v_n), r, z),$$

where

$$r_j = g_j \ \text{ for } \ 1 \leq j \leq n.$$

Since $J_0 \in D$, we have by (4) and by the definition of J_0,

$$z \leq \inf\{\mathscr{A}_i(P_\eta)(g_1(i), \ldots, g_n(i)) : i \in J_0\} \leq x,$$

which is a contradiction. Next suppose that

$$x < z < y \text{ for no } z \in X. \tag{5}$$

Let

$$J_1 = I(P_\eta(v_1, \ldots, v_n), r, x).$$

By (5),

$$x < y \leq \inf\{\mathscr{A}_i(P_\eta)(g_1(i), \ldots, g_n(i)) : i \in J_1\},$$

which contradicts the definition of x. So, let us suppose $y < x$. Then there exists $J \in D$ such that

$$y < \inf\{\mathscr{A}_i(P_\eta)(g_1(i), \ldots, g_n(i)) : i \in J\} \leq x.$$

Let z be the middle member of the inequality above. Since $J \in D$ and D is generated by E, there exist $J_1, \ldots, J_n \in E$ such that

$$J_1 \cap \ldots \cap J_n \subset J.$$

We can assume that there exist $r \in R'^{\infty}$, $\varphi_1, \ldots, \varphi_n \in \Lambda$, and $y_1, \ldots, y_n \in X$ such that for $1 \leq j \leq n$,

$$\begin{cases} y_j < \varphi_j[B, h \circ r], \\ J_j = I(\varphi_j, r, y_j), \\ r_j = g_j. \end{cases} \tag{6}$$

For a fixed j, we have, by (6),

$$t\varphi_j[B, h \circ r] < ty_j.$$

We pick $t_j \in \mathscr{H}$ so that

$$t_j t\varphi_j[B, h \circ r] \neq \underset{\sim}{1} \text{ and } t_j ty_j = \underset{\sim}{1}. \tag{7}$$

Let $\varphi_{n+1} = P_\eta(v_1, \ldots, v_n)$ and let $t_{n+1} \in \mathcal{H}$ be such that

(8) $$t_{n+1}y \neq \underset{\sim}{1} \quad \text{and} \quad t_{n+1}z = \underset{\sim}{1}.$$

Finally, let $\varphi = \max(t_1 t\varphi_1, \ldots, t_n t\varphi_n, t_{n+1}\varphi_{n+1})$. Clearly φ is a conditional formula. It is easy to verify that

$$\{i: \varphi[A_i, r^i] = \underset{\sim}{1}\} \supset J \cup \bar{J}_1 \cup \ldots \cup \bar{J}_n \supset I.$$

Hence, $\varphi[B, h \circ r] = \underset{\sim}{1}$. However, this is impossible due to (7) and (8). So we have proved that $x = y$. The lemma is proved.

The next theorem is the main result of this section. It requires the generalized continuum hypothesis.

THEOREM 7.4.23. Assume the generalized continuum hypothesis. Let $K \in EC_\Delta$. Then K is a conditional class if and only if K is closed under reduced products.

PROOF. As we have pointed out earlier, if K is a conditional class, then it follows from Exercise 7K that K is closed under reduced products. So let us assume that K is closed under reduced products. Let

$$\Delta = \{\varphi \in \Gamma: \varphi[A] = \underset{\sim}{1} \text{ for all } A \in K\}.$$

We first observe that since $\psi = (\sup v_0)\, t(v_0 \equiv v_0) \in \Gamma$, if $\Delta = \Gamma$, then $K = 0$ and

$$K = \{A \in \mathcal{M}: \psi[A] = \underset{\sim}{1}\}.$$

So, let us assume that $\Gamma - \Delta \neq 0$. Let

$$L = \{A \in \mathcal{M}: \varphi[A] = \underset{\sim}{1} \text{ for all } \varphi \in \Delta\}.$$

We prove that $K = L$. Clearly $K \subset L$. Let $B \in L$. By the upward Löwenheim-Skolem theorem, we may assume that

(1) either B is finite or B is special of power α^+,

where $\|\mathcal{L}\| \leq \alpha$. For each $\varphi \in \Gamma - \Delta$, we invoke the downward Löwenheim-Skolem theorem and find $A_\varphi \in \mathcal{M}$ such that

(2) $$A_\varphi \in K, \ \varphi[A_\varphi] \neq \underset{\sim}{1}, \ \text{and} \ \|A_\varphi\| \leq \alpha^+.$$

Let $I = (\Gamma - \Delta) \times \alpha$. Whenever $i = \langle \varphi, \nu \rangle \in I$, we let

$$A_i = A_\varphi.$$

Notice that

(3) $|I| = \alpha$ and if $\varphi \in \Gamma$ and $\{i: \varphi[A] = \underset{\sim}{1}\} \in S^\alpha(I)$, then $\varphi[B] = \underset{\sim}{1}$.

Since (1), (2), (3) hold, we may now apply Lemma 7.4.22 and obtain

$$B \cong D\text{-prod}\ \lambda i\ A_i \ \text{for some filter} \ D \ \text{over} \ I.$$

Since K is closed under reduced products, $D\text{-prod}\ \lambda i\ A_i \in K$, and since $K \in EC_\Delta$, $B \in K$. So $L \subset K$ and the theorem is proved.

EXERCISE 7L.* Assume the generalized continuum hypothesis. Let $K \in EC_\Delta$. Prove or disprove the following: K is an intersection of finite unions of classes of the form

$$\{A \in \mathcal{M}: \ \varphi[A] = \underset{\sim}{1}\},$$

where $\varphi \in \Gamma$, if and only if K is closed under reduced powers.

HISTORICAL NOTES

A comprehensive bibliography of papers in model theory can be found in the soon to be published Proc. of Model Theory Symposium, Berkeley.

Chapter I.

1.2 The theorem that every filter can be extended to an ultrafilter is due to Tarski [1930]. Regular and uniform ultrafilters have been considered in earlier papers in model theory, e.g. Morel-Scott-Tarski [1958], Frayne-Morel-Scott [1962]. The notion of a D-product of sets and of models, see Chapter V, goes back to Skolem [1934] in his construction of a non-standard model of arithmetic. More recently, the construction was applied by Hewitt [1948] to real closed fields, and, more generally, by Łoś [1955]. The present formulation of D-products of sets, as well as Lemmas 1.2.1, 1.2.2, and Exercise 1E, 1F, 1G, 1H, are essentially in Frayne-Morel-Scott [1962], where a more complete historical account may also be found. Exercises 1B, 1C, 1I, 1K are special cases of results in Keisler [1964b].

1.3. The topology $\mathcal{S}^*$ over X^* is due to Vietoris [1923].

1.5. The D-limit is closely related to Moore-Smith convergence (see Kelley [1955]) and to convergence of filters (see Bourbaki [1953]). For a brief discussion of the connection of D-limits with Moore-Smith convergence, see Hildebrandt [1963]. Theorem 1.5.8 is due to Vietoris [1923].

Chapter II.

2.3. For a comprehensive account of two valued logic, see Church [1956]. The notions introduced in 2.1 and 2.2 are suggested by the corresponding things in two-valued logic.

2.5. Example 2.5.2 goes back to Łukasiewicz [1930]. For the development of the syntax and model theory for this logic, see Wajsberg [1935], Rose-Rosser [1958], Chang [1959a], Rutledge [1959], Chang [1961], Scarpellini [1962], Belluce-Chang [1963], Hay [1963], and Belluce [1964]. Some applications of this logic are found in Skolem [1957], Scarpellini [1963], Chang [1963a], [1965]. Some papers and books containing results on many-valued logics with other value spaces are: Post [1921], Rosser-Turquette [1952], Kleene [1952], Mostowski [1961], Chang-Keisler [1962], Mostowski [1963], Chang [1963], Rasiowa-Sikorski [1963], Mostowski [1964], and Chang-Keisler [1966].

The Lindenbaum algebra mentioned in Exercise 2D can be found in Tarski [1956]. Brouwerian algebras and the interpretation of intuitionist logic are discussed, for example, in Rasiowa-Sikorski [1963].

Chapter III.

Most of the basic notions of model theory for the classical logic ℓ are due to Tarski [1935], [1956]. The intensive development of this subject began with the papers of Henkin [1949], Robinson [1951], and Tarski [1952]. The method of extending a model A to a model (A, a) is called the method of diagrams and was introduced by Robinson [1951] and Henkin [1953].

Our notion of a closed theory corresponds to the usual notion of a (deductively closed) theory in two-valued logic; our notion of a theory corresponds to a set of complete theories in two-valued logic.

IN THE NOTES ON THE REMAINING CHAPTERS, WHEN WE REFER TO A RESULT OR TO AN EXERCISE, WE HAVE IN MIND THEIR SPECIALIZATION TO TWO-VALUED LOGIC ONLY.

Chapter IV.

4.1. Extended theories of models A of two-valued logic can be thought of as sets of complete types of elements of A; they have been considered, for example, in Vaught [1961], Morley [1963].

4.2. Elementary extensions of models are due to Tarski, see Tarski-Vaught [1957]; most of our results in 4.2 and 4.3 are generalizations of results in Tarski-Vaught [1957]. The downward Löwenheim-Skolem theorem found in Tarski-Vaught [1957] is an improvement of a theorem of Skolem [1920], which is in turn an improvement of a theorem of Löwenheim [1915]. Löwenheim's theorem is generally considered to be the earliest result in model theory.

Chapter V.

5.1. The fundamental lemma (Lemma 5.1.4) is stated by Łoś [1955] and a proof can be found in Frayne-Morel-Scott [1962]. Exercise 5C is due to Frayne and is also in Frayne-Morel-Scott [1962].

5.2 The compactness theorem (Theorem 5.2.2) is due to Gödel [1930] for countable similarity types and to Malcev [1936] for uncountable similarity types. The proof using ultraproducts is due to Morel-Scott-Tarski [1958] and is given in Frayne-Morel-Scott [1962]. Versions of the compactness theorem for the value space [0, 1] can be found in Chang [1961]; for compact Hausdorff spaces in Chang-Keisler [1962], [1966]; Mostowski [1963] gave a proof, without using ultraproducts, of the theorem in Chang-Keisler [1962].

5.3. Theorem 5.3.1 is given in Frayne-Morel-Scott [1962]; so are Exercises 5G, 5H, 5I.1. The history of the upward Löwenheim-Skolem theorem is discussed in Vaught [1954], see also Tarski-Vaught [1957]. Exercise 5K can be found in Vaught [1954] and in Łoś [1955a]. Exercise 5L is based on an observation of Hanf [1962]. The number β in Exercise 5L is sometimes called the Hanf number for the logic.

5.4. The notions and results of this section are taken from Keisler [1964a].

5.5. Theorem 5.5.4 is due to Keisler [1964].

Chapter VI.

6.1. For a discussion of the origins of the notion of

α-saturated models, which is rather complicated, see Morley-Vaught [1962], where they consider the closely related homogeneous universal models (see also Jónsson [1960]). Theorems 6.1.7 and 6.1.9 are essentially in Morley-Vaught [1962].

6.2. Special models were first introduced in Morley-Vaught [1962]. Our definition of a special model is equivalent to the one given in Morley-Vaught [1962], but is simpler. Theorem 6.2.9 is in Morley-Vaught [1962].

6.3. See Morley-Vaught [1962] also for the notions of universal models. Our development in this section is very close in spirit to the development found there.

6.4. Theorem 6.4.1 is given in Morley-Vaught [1962]. Exercises 6P and 6Q are closely related to a characterization of elementary classes using limit ultrapowers, see e.g. Keisler [1963]. Exercise 6R is essentially a theorem of Specker [1962]. Exercise 6S is Robinson's consistency theorem [1956]. Exercise 6T is the Craig interpolation theorem [1957]. An earlier paper of Beth [1953] is also relevant.

6.5. This section parallels a development of Keisler [1961]. Theorem 6.5.1 for the value space [0, 1] is stated in Chang [1961], and for continuous logics in Chang-Keisler [1962].

Chapter VII.

References to the principal results of this chapter are given in the beginning of the chapter. Several of these results for the value space [0, 1] are stated in Chang [1961].

7.2. Various parts of Exercise 7F are proved in Łoś-Suszko [1957], Chang [1959], Robinson [1959], and Keisler [1960]. Exercise 7I can be found in Keisler [1960].

7.4. The notion of reduced products can be found in Łoś [1955] and Frayne-Morel-Scott [1962]. A very special form of this construction was used in Chang-Morel [1958]. Horn [1951] first realized that conditional sentences are preserved under direct products; Chang saw that this is still true for reduced products. Chang-Morel [1958] showed that not all sentences preserved under direct products are equivalent to conditional sentences.

BIBLIOGRAPHY

Belluce, L. P.

1964. Further results on infinite valued predicate logic, J. Symbolic Logic 29, 69-78.

Belluce, L. P. and Chang, C. C.

1963. A weak completeness theorem for infinite valued predicate logic, J. Symbolic Logic 28, 43-50.

Beth, E. W.

1953. On padoa's method in the theory of definition, Indag. Math., 56, 330-339.

Bourbaki, N.

1953. Élements de Mathematiques XVI Pt. I, livre III: Topologie genérale, Actualités sci. et ind., No. 1196.

Chang, C. C.

1959. On unions of chains of models, Proc. Amer. Math. Soc., 10, 120-127.

1959a. A new proof of the completeness of the Łukasiewicz axioms, Trans. Amer. Math. Soc., 93, 74-80.

1961. Theory of models of infinite valued logics, I-IV, abstracts in the Notices of Amer. Math. Soc., 8, 68 and 141.

1963. Logic with positive and negative truth values, Modal and many-valued logics, Acta Philosophica Fennica, Fasc. XVI, 19-39.

1963a. The axiom of comprehension in finite valued logic, Math. Scand., 13, 9-30.

1965 Infinite valued logic as a basis for set theory, Proc. of the 1964 Int'l Congress for Logic, Methodology, and Philosophy of Science, Jerusalem, Israel, 93-100.

Chang, C. C. and Keisler, H. J.

1962. Model theories with truth values in a uniform space, Bull. Amer. Math. Soc., 68, 107-109.

1966. Continuous model theory, to appear in Proc. of Model Theory Symposium, Berkeley.

Chang, C. C. and Morel, A. C.

1958. On closure under direct product, J. Symbolic Logic, 23, 149-153.

Church, A.

1956. Introduction to mathematical logic, volume 1, Princeton University Press.

Craig, W.

1957. Three uses of the Herbrand-Gentzen theorem in relating model theory and proof theory, J. Symbolic Logic, 22, 269-285.

Frankel, H.

1964. On the fringe: Botany, Barometers, and Bridgework, Saturday Review, vol. 47, no. 22, (May 30), 31 and 52.

Frayne, T., Morel, A. C., and Scott, D.

1962. Reduced direct products, Fund. Math., 51, 195-228.

Gödel, K.

1930. Die Voltständigkeit der Axiome des logischen Funktionenkalküls, Monatsh. und Physik, 37, 439-360.

Hanf, W.

1962. Some fundamental problems concerning languages with infinitely long expressions, Ph.D. thesis, University of California, Berkeley.

Hay, L. S.

1963. Axiomatization of the infinite-valued predicate calculus, J. Symbolic Logic, 28, 77-86.

Henkin, L.

1949. The completeness of the first-order functional calculus, J. Symbolic Logic, 14, 195-166.

1953. Some interconnection between modern algebra and mathematical logic, Trans. Amer. Math. Soc., 74, 410-427.

Hewitt, E.

1948. Rings of real-valued continuous functions, Trans. Amer. Math. Soc., 64, 45-99.

Hildebrandt, T. H.

1963. Introduction to the theory of integration, Academic Press, New York.

Horn, A.

1951. On sentences which are true of direct unions of algebras, J. Symbolic Logic, 16, 14-21.

Jónsson, B.

1960. Homogeneous universal relational systems, Math. Scand., 8, 137-142.

Keisler, H. J.

1960. Theory of models with generalized atomic formulas, J. Symbolic Logic, 25, 1-26.

1961. Ultraproducts and elementary classes, Indag. Math., 23, 477-495

1963. Limit ultrapowers, Trans. Amer. Math. Soc., 107, 382-408.

1964. Ultraproducts and saturated models, Indag. Math., 26, 178-186.

1964a. Good ideals in fields of sets, Ann. Math., 79, 338-359.

1964b. On cardinalities of ultraproducts, Bull. Amer. Math. Soc., 70, 643-646.

1965. Reduced products and Horn classes, Trans. Amer. Math. Soc., 117, 307-328.

1965a. A survey of ultraproducts, Proc. of the 1964 Int'l Congress for Logic, Methodology, and Philosophy of Science, Jerusalem, Israel, 112-126.

Kelley, J. L.

1955. General topology, Van Nostrand, Princeton, N. J.

Kleene, S. C.

1952. Introduction to metamathematics, Van Nostrand, Princeton, N.J.

Łoś, J.

1955. Quelques remarques, théorèmes, et problèmes sur les classes définissables d'algebres, Mathematical interpretation of formal systems, Amsterdam, 98-113.

1955a. On the categoricity in power of elementary deductive systems and some related problems, Coll. Math., 3, 58-62.

1955b. On the extending of models, I, Fund. Math., 42, 38-54.

Łoś, J. and Suszko, R.

1957. On the extending of models, IV, Fund. Math., 44, 52-60.

Löwenheim, L.

1915. Über Möglichkeiten in Relativkalkül, Math. Ann. 76, 447-470.

Łukasiewicz, J. and Tarski, A.

1930. Untersuchungen über den Aussageukalkül, C. R. des Séances de la Société des Sciences et des Lettres de Varsovie, Classe III, 23, 30-50.

Lyndon, R. C.

1959. Properties preserved under homomorphism, Pacific J. Math., 9, 143-154.

Malcev, A.

1936. Untersuchungen aus dem Gebiet der mathematischen Logik, Mat. Sbornik (2), 323-336.

Morel, A. C., Scott, D., and Tarski, A.

1958. Reduced products and the compactness theorem, Notices Amer. Math. Soc., 5, 674-675.

Morley, M.

1963. On theories categorical in uncountable powers, Proc. Nat. Acad. Sci., 49, 213-216.

Morley, M. and Vaught, R.

1962. Homogeneous universal models, Math. Scand., 11, 37-57.

Mostowski, A.

1961. Axiomatizability of some many valued predicate calculi, Fund. Math., 50, 165-190.

1963. The Hilbert epsilon function in many-valued logics, Modal and many-valued logics, Acta Philosophica Fennica, Fasc. XVI, 169-188.

1964. An example of a non-axiomatizable many-valued logic, z. Math. Logik, 7, 72-76.

Post, E. L.

1921. Introduction to a general theory of elementary propositions, Amer. J. Math., 43, 163-185.

Rasiowa, H. and Siborski, R.

1963. The mathematics of metamathematics, Monografie Matematyczne, Warsaw.

Robinson, A.

1951. On the metamathematics of algebra, Amsterdam.

1956. A result on consistency and its applications to the theory of definition, Indag. Math., 18, 47-58.

1959. Obstructions to arithmetical extensions and the theorem of Łoś and Suszko, Indag. Math., 21, 489-495.

1963. Introduction to model theory and to the metamathematics of algebra, Amsterdam.

Rose, A. and Rosser, J. B.

1958. Fragments of many-valued statement calculi, Trans Amer. Math. Soc., 87, 1-53.

Rosser, J. B. and Turquette, A. R.

1952. Many-valued logics, Amsterdam.

Rutledge, J. D.

1959. A preliminary investigation of the infinitely many-valued predicate calculus, Ph.D. thesis, Cornell.

Scarpellini, B.

1962. Die Nichtaxiomatisierbarkeit des unendlichwertigen Prädikatenkalküls von Łukasiewicz, J. Symbolic Logic, 27, 159-170.

1963. Eine Anwendung der unendlichwertigen Logik auf topologische Räume, Fund. Math., 52, 129-150.

Skolem, T.

1920. Logisch-kombinatorische Untersuchungen über die Erfüllbarkeit oder Beweisbarkeit mathematische Sätze nebst einem Theoreme über dichte Mengen, Skrifter utgit av Videnskapsselskapet i Kristiania, I. Matematisk-naturvidenskabelig klasse 1920, No. 4.

1934. Über die Nicht-Charakterisierbarkeit der Zahlenreihe mittels endlich oder abzählbar unendlich vieler Aussagen mit ausschliesslic Zahlenvariablen, Fund. Math., 23, 150-161.

Specker, E.

1962. Typical ambiguity, Proc. of 1960 Int'l Congress for Logic, Methodology, and Philosophy of Science, 116-124.

Tarski, A.

1930. Une contribution à la theorie de la mesure, Fund. Math., 15, 14-20.

1935. Grundzüge des Systemkalküls, I and II, Fund. Math., 25, 503-526, and 26, 283-300.

1952. Some notions and methods on the borderline of algebra and metamathematics, Proc. 1950 Int'l Congress of Mathematicians, 705-720.

1954. Contributions to the theory of models, Indag. Math., 16, 572-588.

1956. Logic, Semantics, Metamathematics, Oxford.

Tarski, A. and Vaught, R.

1957. Arithmetical extensions of relational system, Compositio Math., 13, 81-102.

Vaught, R.

1954. Applications of the Löwenheim-Skolem-Tarski theorem to problems of completeness and decidability, Indag. Math., 16, 467-472.

1961. Denumerable models of complete theories, Infinitistic Methods, Warsaw, 303-321.

1963. Models of complete theories, Bull. Amer. Math. Soc., 69, 299-313.

Vietoris, L.

1923. Bereihe Zweiter Ordnung, Monatsh. Math. Physik., 33, 49-62.

Wajsberg, M.

1935. Beiträge zum Metaaussagenkalkül I, Monatsh. Math. Physik., 42, 221-242.

INDEX OF SYMBOLS

INDEX OF DEFINITIONS

INDEX OF EXERCISES

Ingram Content Group UK Ltd.
Milton Keynes UK
UKHW011243290523
422449UK00001B/70

9 780691 079295